KU-516-163

Science

The Encyclopedia of
THE PLANT KINGDOM

Chamaemelum nobile
Chamomile

THE BRITISH COUNCIL LIBRARY
NEW DELHI

EXHIBITION COPY
NOT FOR SALE

The Encyclopedia of
THE
PLANT
KINGDOM

a Salamander book

Published by
HAMLYN
London · New York · Sydney · Toronto

A Salamander Book

This edition published 1977 by
The Hamlyn Publishing Group Limited
London · New York · Sydney · Toronto
Astronaut House, Feltham,
Middlesex, England.

ISBN 0 600 33134 2

© Salamander Books Ltd 1977
52 James Street
London W1
United Kingdom

All rights reserved. No part of this book may
be reproduced, stored in a retrieval system
or transmitted in any form or by any means,
electronic, mechanical, photocopying,
recording or otherwise, without the prior
permission of Salamander Books Limited.

Filmset by
SX Composing, Leigh-on-Sea,
Essex, England

Colour reproduction by
Paramount Litho, Basildon, Essex, England

Printed by Henri Proost,
Turnhout, Belgium

All correspondence concerning the content of
this volume should be addressed to
Salamander Books Limited

Acknowledgements

The publishers wish to express their sincere
thanks to **Heather Angel** for supplying the
majority of the excellent colour photographs in
this book; her efficiency and enthusiasm helped
greatly towards its trouble-free production.
Heather Angel travels extensively in search of
new subjects. Using her biological background,
combined with a perceptive pictorial apprecia-
tion, she is able to create authentic and exciting
images of nature. Her cameras are used for any
subject—no matter what size—but it is the
close-up world which is her forte. Heather is
the author of six books on the techniques of
nature photography, as well as many more on
wildlife topics.

We also owe thanks to **Dr Pat Morris** and
Dr Glyn Jones for their radical and lively plan
of the book, and to **Chris Steer** as designer.

The Authors

Hilary Bladon studied botany and zoology at Royal Holloway College (London University) and gained a first class honours degree in these subjects. She now lives in Dorset and concentrates on writing.

Dr J. M. Chapman gained a first class degree with honours in chemistry at the Imperial College of Science and Technology, London, and subsequently won a Nuffield Scholarship to Queen Elizabeth College. Here he researched in plant physiology to gain his doctorate and since 1967 has lectured at the college.

Dr Peter Crisp has a degree in botany from Queen Mary College, London, and a doctorate in Experimental Taxonomy. He has worked at the National Vegetable Research Station, Wellesbourne, Warwickshire, since 1969, where he is engaged in the breeding of vegetables (particularly cauliflower) to produce a predictable and uniform yield.

D. P. Gibson gained a degree with honours in botany from St Andrews University, Fife, and won a diploma in Plant Taxonomy from the University of Edinburgh. He is currently working towards a doctoral thesis on oceanic plants, and has been a member of several expeditions to Greenland, France and the Canary Islands, where he has collected and studied plants.

Dr Barry Giles was educated in Guildford, Surrey, and studied botany at Westfield College, London, both as an undergraduate and postgraduate. His main interest is research into sphagnum bogs and he is preparing a thesis on this subject for his doctorate. He is senior lecturer in biology at the City of London Polytechnic, specialising in ecological studies.

Herbert L. Edlin, has worked at the Forestry Commission as Publications Officer since 1945 after a varied career in agriculture, including managing a rubber estate in Malaysia. He read botany and forestry at Edinburgh and Oxford Universities, and is the author of many plant books.

Hannah Grimes studied botany at the University of Aberystwyth, Wales, and after obtaining an honours degree undertook two years of postgraduate research into ecology. She now writes in her spare time.

J. W. Grimes studied botany at Manchester University at undergraduate and postgraduate levels. He is now employed as a fern taxonomist in the herbarium of the Royal Botanic Gardens, Kew.

Philip Horton was educated in Bath and graduated with an honours degree in botany from University College, London, in 1969. He is now professionally involved in nature conservation in Wiltshire.

Dr John Mason studied zoology and botany at University College, London, and obtained an honours degree in zoology. He gained his doctorate with a thesis on the ecology of woodland beetles and since 1970 has worked for what is now called the Nature Conservancy Council. As Deputy Regional Officer he is mainly responsible for advising other departments and landowners on the practical conservation of wildlife and for running wildlife reserves which the Council has under its control.

Allen Paterson studied at the University Botanic Garden, Cambridge, and at the Royal Botanic Gardens, Kew. After taking a Certificate in Education he taught Rural and Environmental Studies in schools and colleges of education for 14 years. He is now Curator of the Chelsea Physic Garden and writes on horticultural subjects.

Dr Frank Slater studied botany at the University College of Wales, Aberystwyth, and gained an honours degree in science. After some teaching and research on pollen analysis he returned to Aberystwyth in 1970 to teach botany. He gained his doctorate with a thesis on the ecology of peat lands, and has been with the University of Wales Institute of Science and Technology since 1974. His current position is as curator of the Institute's Llysdinam Field Centre near Llandrindod Wells.

A. H. M. Synge studied horticulture at Wye College, University of London. He is now research assistant for the Threatened Plants Committee of the International Union for Conservation of Nature and Natural Resources (IUCN), based at The Royal Botanic Gardens, Kew. He is also involved in compiling the IUCN "Red Data Book" for plants. He has undertaken expeditions to Malawi, looking at plants and national parks, and travelled in the USA, Russia and much of Europe, particularly Greece to study plants. Mr Synge would like to thank in particular **Dr J. Dransfield, Mr G. Ll. Lucas** and **Dr R. Melville** for their help and for the use of unpublished information, and to acknowledge that the section of rain forest, an area he has not visited, is heavily based on the literature given in the select bibliography which appears on page 240.

Botanical Editor

Anthony Huxley is an experienced writer with 23 publications already to his name. He has travelled widely to study and photograph plants in the wild, and when at home is a keen gardener. He worked on the weekly publication *Amateur Gardener* for over 20 years, becoming its editor in 1967. He relinquished this post in 1971 to devote himself to serious writing.

Introduction

by Anthony Huxley

The plant world is often taken very much for granted—as just so much greenery. Plants seem part of the landscape, like hills or houses, not living breathing organisms each with a life of its own. When we walk on grass, how many of us stop to think of the complexity of life in every single cell crushed beneath our feet? But this green world which surrounds us is supremely important to mankind since all creatures depend in the first place upon plants. Even carnivorous animals rely on eating other animals which in turn have fed on plants. The renewal of the earth's oxygen supply is a vital service performed silently and unseen by green plants.

Plants form a fascinating world of their own, where shell, bone, muscle, blood and nerves are replaced by cellulose, wood, sap and chemical communications. It is a world actuated by chlorophyll, the unique substance which allows plants to use the energy of sunlight to manufacture food. Long before man evolved, plants had solved all sorts of problems from structural engineering to plumbing, had mastered hydrodynamics and aerodynamics, and had 'invented' a host of ingenious gadgets.

Plants, like animals, have powers of movement and feeling, but the means and consequences differ. Their methods of growth, reproduction and dissemination are different, yet serve similar ends, often with a fascinating diversity of mechanisms.

Part one of this encyclopedia outlines how plants grow, work, reproduce and spread, how they are adapted to occupy almost any niche on the globe and form balanced communities of varying complexity suitable for the different habitats earth provides.

Part two describes some major plant groups. Many of these are natural groups, while others relate to man's uses and general classification of plants. In his use of plants man has often replaced natural species with those he has bred, far removed from their natural ancestors. Both to make room for these artificial offspring, and equally to exploit the results of millions of years of evolutionary development, he is gradually destroying more and more great areas of natural plant growth. Worldwide, over 20 hectares (50 acres) of vegetation are being destroyed every *minute* (11 million hectares [28 million acres] a year) and a chapter is devoted to this terrifying problem.

Man is in a hurry, and in a panic too as his numbers increase in geometric progression and threaten to outstrip the earth's capacity to support him. In his panic he fails to remember what sudden and clumsy action inevitably does to the delicate balance of plants, animals and terrain.

We who are alive today are fortunate in still being able to see a range of fascinating plants and natural communities. This encyclopedia offers an insight into their distinct, universal and remarkable world. Hopefully this insight will lead to a better understanding of man and his battered, overcrowded planet and of the fundamental role of the plants that share it with us.

Anthony Huxley

Contents

THE
LIVING WORLD
OF PLANTS

Viola odorata
Sweet Violet

THE LIFE OF PLANTS

"Splendour in the grass, glory in the flower"

Colchicum autumnale
Meadow Saffron or Autumn Crocus

There can be few so commonly misquoted quotes which are so universal, so vital and so imperfectly understood as the famous 'all flesh is as grass'. The evangelist was commenting, no doubt, upon man's ephemeral nature—as tenuous as, in another context, are 'the flowers of the field'. Such early references to the world of plants are legion in the writings of all cultures reflecting man's consciousness of the distaff side of animate things. Yet in leaving behind his early beginnings, in developing a hierarchical and highly differentiated series of civilizations, innate consciousness that all flesh *is* grass has not been at all generally replaced by knowledge of the scientific truths that make it more valid now, in a period of frighteningly rapid population increase, than ever in the past.

Plants, and plants only, possess the amazing ability to make, synthesize, produce—indeed create, out of the very air—the complicated foods that are the source of the plants' own energy and the source, too, of animal life as we know it on this planet.

Such a claim may appear a fulsome hybrid between the mediaeval elixir of life and a science fiction parody. Yet it is easily, if not simply justified. The omnipresent ingredients of gaseous carbon dioxide and of water, with sunlight as the energy-source are built up into starches and sugars which are used in a wide range of forms as the basic growth material of plant bodies. And plant bodies are the basic growth material in the unbroken organic chain that stretches from amoeba to man. The process is called photosynthesis. But why plants? The simple answer is their possession of what became so verbally debased when the toothpaste and deodorant manufacturers got hold of the name—'nature's green'—chlorophyll.

In the cells of plant leaves and other green parts are small organs, that are unique to plants, called chloroplasts and which control the chemical reaction that is continually taking place in the presence of light (even artificial light). The essential ingredient, chlorophyll, acts as a chemical catalyst—a facilitator which remains itself unchanged. All normal wavebands of light can be used except green: this, being reflected, causes us to 'see' plants as green and to take such greenness as the essential character of the plant kingdom, which both visually and chemically, it actually is.

This basic characteristic—the possession of chlorophyll—is that which defines a plant except in a few very specialized cases. Conversely (with a few exceptions) any living organism that is without them must be an animal. Conventional definitions that members of the animal kingdom are characterized by movement are altogether too simplistic: plants move, in response to a range of stimuli, as well. No, the difference is that animals cannot photosynthesize; they have no chlorophyll. Hence, in truth, all flesh is grass.

It is, of course, unhelpful merely to refer to the plant kingdom as a unit because of its one great universal denominator. The diversity of form is enormous from the mass of microscopic, often unicellular, plants which make up (with a world of equally microscopic animicules) the green scum in pondwater, to trees almost as high as the spire of Salisbury Cathedral or as old as the Ark—and still living. Such diversity, which is, after all, the response to environment throughout evolutionary aeons, demonstrates the all-pervading success story of plants. There is, and as long as life has existed on this planet, there always has been, a plant for all seasons and for all sites.

Classification of plants

As with all life it seems likely that plants began as waterborne organisms. Changes in environment caused the forms to spread to dry land and to colonize all but the most inhospitable terrain. And from land further forms have moved back into the water. The processes of adaptation, modification and opportunism have been and are continuous. Whenever man, since classical times, has started to study the world of nature he has had to try to classify, to list, to order, as much for his peace of mind as for the facilitation of the study. Any discussion of this truly daunting diversity is bound to do the same—although, of course, as time goes on and more work is done—the classification used becomes more sophisticated. Consensus is still not reached.

An accepted scheme is to divide the plant kingdom up into seventeen divisions, each sub-divided into classes, subclasses, orders, tribes and families before it is possible to get down to actually naming a plant. Such a classification aims to do two main jobs in addition to making conversation possible about them. These are to suggest relationships and to indicate an evolutionary sequence: these roles, of course, are inter-related. Only the final two divisions consist of flowering plants—the flowering process being the height of plant development.

In looking, then, at the world of plants and trying to understand their ordering, it is reasonable to start at the simplest and hence, in all probability, the most primitive in evolutionary terms. At the beginning come the bacteria, the first of the seventeen divisions. Bacteria, which are probably on the borderline between plants and animals, have no chloro-

Below: Sheep eat grass; men eat sheep—without plants no animals could exist, for plants can use the sun's energy to create food from air and water alone.

Above: Lowly and advanced plants live together in every kind of niche on the earth's surface. Here a filamentous alga forms "scum" in water on which float the aquatic fern Azolla and duckweed, a flowering plant.

Left: Great conifers like these giant redwoods, tough and long lived, are seed-bearing plants which have existed on earth for over 200 million years, long before the flowering plants, and still grow in most parts of the globe.

phyll, no clear nucleus in their unicellular bodies and reproduction is by simple division.

The succeeding seven divisions comprise a world in themselves, generally unknown and unregarded but fulfilling their own diverse roles species by species and playing their part in the great multidimensional web of organic life. These are the algae. They vary in size from the microscopic to the massive. They include the extraordinary diversity of the seaweeds, the biggest of which can grow up to 90m (300ft) in length in tropical seas. As non-flowering plants, algae frequently reproduce by simple fragmentation of their vegetative bodies but a sexual stage is also usual (as, for example, the 'conjugation' of spirogyra).

Their success in adapting to their environment should need no emphasis though the potential difficulties might be considered. There is no necessity for aquatic species to be able to abstract carbon dioxide and oxygen from the water; there may be greatly reduced light and in addition marine plants have to deal with the problem of living in a highly concentrated solution of salts where 'normal' plants could never survive.

The division called the Mycophyta is a difficult one to place. These are the fungi which do not photosynthesize and, possessing no chloroplasts, are not green (the brown and red seaweeds have pigments which mask their chlorophyll). Fungi have taken a cuckoo-like path to success and exist as two types: as saprophytes which live on the dead remains of organic material as do mushrooms and moulds on old bread. Other moulds and mildews are crippling diseases of higher plants and hence are parasites. Some parasites are said to be faculta-tive: having attacked and killed their host, they then continue to live on its decaying body—a gory sequence in anthropomorphic terms. Not needing to photosynthesize, fungi do not need light for growth and hence mushrooms can be cultivated in cellars.

Lichens, whose growth on walls and roofs helps man's artifacts to blend in with their surroundings, are odd organisms in that each species is a combination of an alga and a fungus which pair symbiotically to exist in an utterly unlikely marriage. In areas of high humidity large lichens of bizarre form festoon trees and cover rocks. (Habitat and environment are apt to encourage similarity of form in widely differing plants. The famous Spanish moss of the Florida Everglades which so resembles a big grey lichen, is neither this, nor a moss, but a flowering plant of the pineapple (Bromeliaceae) family. To assume relationships from general appearances is hazardous.)

The next division in the plant kingdom is the Bryophyta, which are separated into the mosses and liverworts. The use of the plural should be noted. So often we talk of having 'moss in the lawn' or seeing liverwort by the riverbank without appreciating that these are just two of a huge number of species in a diversity of forms. Most are lowly and flourish in damp spots, often in positions of poor light intensity. They are thus adapted to succeed as the lowest layer in, for example, a woodland stratification.

Mosses are usually leafy in form (especially when examined under a lens) whilst most liver-worts consist of a flat plate of tissue, only a few cells thick, which bears the separate male and female reproductive organs. Much reproduc-tion, however, is by simple fragmentation of the plant itself. Almost any piece broken off will continue to put out rhizoids (primitive roots) if in a suitable spot.

It is not until the Pteridophyta are reached (evolutionarily speaking) that we generally start to notice, and hence accept, clear dif-ferences between one plant and another. True, we are wise to distinguish, in the fungi, between mushroom and deathcap; but mostly we lump everything fungal together as 'toadstools', just as all marine algae become 'seaweed'.

The Pteridophyta includes the ferns whose range of morphological form is not only great but of such beauty to the human eye as to be deemed worthy of cultivation. At this point the gardener enters and we really start to take notice of what the plant kingdom offers.

Ferns, of course, vary in size from tiny plants of rock crevices to 'trees' in tropical jungles. Tree-ferns, although having apparent trunks up to 10m (33ft) high, have not really grown wood as we know it; this is the prerogative of higher groups. The pteridophytes also include various fern-allies of which the club-mosses, the selaginellas and the horsetails (*Equisetum* species) are most distinctive.

The true ferns have a peculiar life-cycle, or, one might say, life-tandem which clearly indi-cates their early origins. Everyone will have noticed the dark dots or bars on the underside of fern fronds producing dust-like spores. These spores will germinate, not into a parent-resembling fern plant, but into a small green thallus like a liverwort. And like a liverwort this will bear sexual organs, the fusion of whose gametes (or sexual cells) will cause the develop-ment of a new recognizable fern plant. This unique sequence is known as the alternation of generations—sporophyte to gametophyte back to sporophyte.

All the great divisions so far mentioned depend very greatly upon water; not just as an ingredient of the photosynthetic process or as that which the 'roots' take up from the soil. High humidity, if not a full watery medium, is essential to life. Otherwise dessication and death quickly sets in. This reverts again, therefore, to the habitat of most of these early groups, the non-flowering plants and to their, sequentially, primitive status.

Any informed knowledge of evolution (the process by which it is maintained that the myriads of plant and animal forms were not individual creations but a continuum of de-velopment in response to habitat and environ-mental change) is likely to be based, or at least helped, by fossil records. Unfortunately, yet inevitably, the most primitive organisms, be-cause of their very nature, do not fossilize. Yet it is well established that the flowering and seeding process which we consider to be so typical of plant life is a late, and, it would seem, cumulative development. Of the number of species which must have existed and are now extinct and even of the current sum of the world's flora the seed-bearing plants must be seen numerically, to have a small place.

However, their success in variation in form, in size and in adaptation makes them para-mount. These seed-bearing plants comprise just two divisions in the seventeen of the plant kingdom. The first, and more primitive, are the Coniferophyta (often called Gymnosperms: literally 'naked seeds'). These include a few small palm-like tropical trees, the cycads, but mainly the conifers, which, as their name sug-gests, carry their seeds on woody cones familiar on pine, cedar, spruce, hemlock, larch and so on. Conifers are all trees, or at least dense woody shrubs, with a worldwide distribution.

Their evolutionary age has been con-veniently shown by fossil records of species still existing. Scientific excitement, comparable

with that which zoologists enjoyed when a coelacanth was fished up alive off Madagascar, was intense when seeds of *Metasequoia glyptostroboides* reached Europe from Hupeh in 1948. The tree had hitherto only been known from Carboniferous fossils over 200 000 000 years old.

The higher flowering plants comprise the last division, the Magnoliophyta (or Angiosperms). They begin, evolutionarily speaking, with the beautiful magnolias (fossil records of magnolia relatives are frequent) and progress through all forms, in all habitats to virtually cover the earth. The life-style of each reflects its own particular success-story (recall Darwin's maxim of the survival of the fittest) and hence they may be woody or herbaceous, aquatics or desert plants, annual or living for millenia and every conceivable intermediate stage.

Basic structure and life cycles of plants

However apparently different in form or feature representatives of the higher plant divisions may be, in function they are remarkably similar. The universality of photosynthesis has already been emphasized; equally important are the processes of respiration and transpiration. These three basic areas are concerned with the exchange of gases from within the plant to the atmosphere (or vice versa) and, combined with other needs, determine plant form.

Basic necessities for plant growth are light, water, air, warmth and nutrients. While all are essential the differences in the availability of each combine to vary environment and, in turn, the plants that grow in any one place. Certain species therefore are more likely than others to inhabit a particular habitat but, in addition, unrelated plants of a distinctive (or restrictive) habitat are apt to exhibit similarities of form. This phenomenon of 'parallel evolution' helps to indicate how plants develop in response to the urge to be among 'the fittest'. A remarkable example is the virtual identical appearance of certain New World cacti to species of African spurges which live in similar habitats. Only their flowers, which utterly betray their true relationships, can be seen to differ completely.

Whether adapted to live in a pond, 5,000m (16,400ft) up on Mount Kilimanjaro, in a desert or a tropical jungle, the basic parts of plants—their essential body—remains the same. The four vital components are: the roots, the stem, its leaves and the flowers. From the latter are the seeds developed and their successful production to ensure the continuation of the species is the entire *raison d'etre* of each and every plant. Plants do not grow to provide food for browsing animals and man, nor come out in glorious flowers for the benefit of bees or to satisfy man's aesthetic sense. They do it for themselves utilizing anything or anybody to further those ends.

The basic parts and their myriad minor and major adaptions are miracles of effectiveness. They have to be: there is always another plant

Left: Fungi may be called plant renegades, for they have abandoned photosynthesis. They are essential to earth's total life-cycle for they reduce once-living material to its basic components, like this bracket fungus finishing off a tree.

only too willing, and usually able, to take up an unused ecological niche.

The first thing to emerge from almost any germinating seed is a root. It has two jobs. First it anchors the growing plant in its place. (Seeds of plants such as mangroves develop their first root like a torpedo whilst still on the parent plant: a small round seed would splash into the water and drift away but a heavy hydrodynamic shape spears through the water and impales itself in the mud beneath ready to get on with the job of living successfully.) Second it is the search for and transport of water and the broad spectrum of nutrients obtained from the soil which are needed to supplement the carbohydrates that the green aerial parts manufacture. All go to building up the plant body, to maintaining it as a healthy organism, to facilitate its reproductive capacities.

The full extent of plants' roots, as they divide, spread and divide again until the unicellular root hairs are reached, is seldom appreciated but it must be expected that the plant will possess as much subsoil growth as that seen in the air. Viewing a tree reflected in water is a good analogy: of course the exact shape will not be repeated—soil texture, rocky-substrate or a high watertable may prevent it but the top will only be as great as the root can support both structurally and physiologically.

After the development of the first root the germinating seed pushes up its first shoot. There may be a single leaf (this monocotyledonous beginning is a valuable diagnostic character of grasses, lilies, palms and a few other plant families) or a pair of seed-leaves (dicotyledonous plants) like two tiny ping-pong bats held on a thin stalk just above ground. At once, being green, they start to take in carbon dioxide, combine it with water that the young root supplies and begin the process of photosynthesis. Like antelopes that must run from the moment of birth, no time must be lost: the struggle, not perhaps to keep head above water, but certainly in the light, has begun.

The stem, which rapidly begins to thicken and grow (and all growth in girth or height is by division and multiplication of the cells, not by any elastic-like stretching) from between the cotyledons (in some plants the cotyledons remain in the seed below ground as inert food stores and the first photosynthetic leaf is a true one) carries unfolding leaves. In the angle of stem and leaf are buds which can grow into side branches to repeat the pattern. Gradually the typical shape of the plant is built up: the speed of growth varies from several metres in one season to the barely measurable and it will, as in all things, be related to habitat.

The stem spreads out its leaves so that they get as much light as necessary, to support structurally a considerable weight and to act as a complicated 'plumbing' system. Nutrients in solution have to be brought up from the roots and carbohydrates, produced in the green parts, have to be moved to where an energy source is required or stored until needed.

In herbaceous plants, whether they be annual or perennial the leafy parts are renewed either entirely from seed or from overwintering groundlevel buds each year. The stems of woody plants which persist, sometimes as vast trunks many hundreds of years old develop specialized methods of annual thickening to ensure preservation of their wood. But the extension growth of tiny twigs at the top of any such giant will be just the same as the first shoot it possessed as it left its seed.

The leaves, whether they belong to chick-weed at ground level or to a great forest tree perform the same functions. They are the power-horses, where the unique photosynthetic ability of plants primarily takes place. It is based upon the exchange of gases which pass in and out of the plant through the stomata—pores usually situated on the underside of plants' leaves.

Monocotyledonous plants, typically, have vertical leaves with stomata on both sides; water-lilies have perforce to site them on the upper surface. Stomata also permit the escape of excess water as vapour, of oxygen as a waste product of photosynthesis in daytime and its ingress at night as a necessity of respiration. Stomata, too, are capable of movement, opening and closing in response to need and climatic conditions.

Yet this is not enough. The need to have some control of water loss, which in severe drought may lead to wilting, dessication or even death, gives the reason for much of the incredible variation in leaf form. Xerophytes (plants naturally adapted to dry habitats) often have silky or white woolliness over their leaves. The pores are thus protected and the pale colour helps to reflect heat. Sometimes stomata are situated in deep depressions. Sometimes, for example in marram grass (the classic sand-dune binder), leaves are rolled into a spill with the pores only on the inside. The variations seem almost infinite, the ultimate being that leaves are dispensed with altogether, as in brooms and gorse in temperate climates and the majority of cacti. Gaseous exchange and photosynthesis therefore has to be possible through stomata on the stems.

Much of the work that plants do, especially those that are likely to live for some years, is concerned with food storage and a whole range of modifications of root, stem or leaf have evolved. Roots may develop swollen tubers as in dahlia, shoots may dive back into the ground and swell into stem-tubers. Potatoes are these (the fact that they have 'eyes', or buds, which grow into further shoots prove that these are in fact stems). Corms of crocus or gladiolus are compressed stems which husband their food-store below ground for a resting season. Bulbs

of daffodil and many other plants act in the same way, but here the resting food-store comprises layers of swollen leaf-bases. These storage organs double as methods of vegetative reproduction. The disadvantage to the plants themselves of their leafy tops or underground food stores is their palatability to herbivorous animals. Man has eaten potatoes (tubers), onions (bulbs), carrots (the overwintering swollen tap-root of a typical biennial plant), taro yams (a horizontal swollen stem or rhizome) since the earliest times. So it behoves plants to defend themselves if they can.

This they do physically with a vast armoury of spines developed from shoots, prickly leaves (if anyone doubts the reason for the barbarity of holly leaves let him note how they become less and less frequent up the tree as danger from browsing animals decreases). Stinging hairs of nettles and other members of the Urticaceae and Loasaceae families are highly sophisticated as are the acrid juices of spurges as repellants to prospective hungry animals. More economical of effort but no less sophisticated are forms which sufficiently resemble the truly unpalatable, such as deadnettles, which are utterly unrelated to true nettles, as to be left alone. All this is necessary if the culmination of the plants' growth pattern is to be reached.

That culmination, of course, is successful sexual reproduction through the flowering and fruiting process. Such success, so vital to the continuation of the species is attended by so many difficulties and dangers that, just as it is the culmination of the plants' efforts so is it the apex of their evolutionary ingenuity. The basic

Right: This Himalayan forest shows how one type of plant will give way to another as the growing conditions alter: deciduous flowering trees fill the lower slopes but cannot withstand the cold of the higher, where conifers take over.

Below: Ferns were the earliest plants to show well developed root and conducting systems. They are very adaptable and widespread, typically enjoying shaded places under trees like these, and man cultivates many of them.

FLOWERS
the supreme achievement

Ever since the first plants with flowers began to grow on the earth, about one hundred million years ago when the higher insects were beginning to appear, they have endured and spread until they assumed their present position as the dominant plants in almost every possible environment. One of the predominant reasons for their success is the great efficiency, compared with other methods, of their reproductive system, which owes much to the activities of various members of the animal kingdom, most particularly the insects, for effecting pollination. The most conspicuous flowers can be thought of as grand 'advertisement' displays, employing sweet nectar or pollen, scent or colour, or, more frequently, a combination of these, as lures for pollinators.

However, some flowers have neither coloured petals, nectar nor scent; often they are small and inconspicuous as well, as in the grasses. These plants rely on other methods of pollination, the need for advertisement disappearing. The pigment chlorophyll, responsible for the green colouration throughout the plant kingdom, is the only colouring agent found in many wind-pollinated flowers—the wind has no eyes for beauty. Such plants have other floral modifications to ensure the effectiveness of their system.

Animal-pollinated flowers show a dazzling variety in the construction of their flowers; no single floral type can provide for all types of potential animal pollinator, so different plants have adapted themselves to pollination by specific groups by offering variations in their flower colour, scent, size, shape, arrangement and type of food available.

Generalized flower structure

Although flowers may show an enormous range in form and be used by taxonomists almost solely to establish distinctions between the smaller groups of the flowering plants, collectively known as the angiosperms, all flowers conform to a common plan. Within this plan the parts that are noticeable usually belong only to the sterile parts of the plant, and the reproductive organs that ensure the continuance of their life cycle are less prominent or even hidden. A look at the structure of a generalized flower will show how each part contributes functionally to the ultimate end—reproduction.

Receptacle

The receptacle is the expanded end of a flower stalk upon which the floral parts are borne in whorls or spirals with very short spaces (internodes) in between. The parts develop on the apex in a similar manner to the leaf development on a vegetative meristem, so that the receptacle corresponds to an apical meristem of a vegetative shoot. However, growth of the receptacle is limited and stops when all the flower parts have been formed. Buds do not form in the axils of the floral parts (at least only rarely) though they invariably do in the axils of foliage leaves. The receptacle varies widely in its shape and size with different species, and may undergo considerable changes with the development of the flower, but its basic role stays the same—that of a firm base and supporting structure for the more showy parts above.

Sepals

The sepals constitute the outermost whorl of the floral parts, the calyx, and may be described as floral leaves. They may be separate, united into a single outer sheath, or in groups of united sepals. Usually green and leaf-like in texture, they have a simple internal structure of loose parenchyma bounded by an epidermal layer on both faces, which are sometimes hairy. The green colour of the sepals comes from chloroplasts in the parenchyma. The sepals protect the developing parts enclosed within while the flower is still a bud. As the flower blossoms, these sepals fold back so the petals can open, sometimes they may even wither and drop off. In some cases the sepals may change, becoming expanded and brightly coloured and appearing

Papaver rhoeas
Corn Poppy

Right: The blooms of the New Zealand buttercup Ranunculus lyallii *look beautiful to us—but in fact all flowers are only leaves, modified in order to attract insect pollinators.*

more like petals, whose purpose they often serve. This is typical of monocotyledons—for example the tulip—where sepals and petals are almost indistinguishable.

Petals

The corolla is the term used to describe the petals of a flower. They are also floral leaves, but have very distinctive characters, being generally brightly coloured, scented, of expanded form and arranged in a showy fashion to attract animal pollinators. As with the calyx, the corolla may be a single unit.

Flower colours are due either to pigments in special living bodies inside the cells known as plastids, or dissolved in the cell-sap. In the former category belong the common carotenes and xanthophylls, collectively known as the carotenoid pigments, giving colours ranging from lemon yellow to tomato red. More than 60 varieties of carotenoid compounds have been isolated and they make up the second largest family of colour in the plant kingdom (green chlorophyll being the first). Anthocyanins belong in the latter category, pigments dissolved in the plants cell sap. They range in shades from palest pink, through vivid reds and blues, to flamboyant purple. Anthoxanthins also belong to this group and give colours from pale ivory to deep yellow. These are both frequently occurring plant pigments.

Sometimes chemical factors can affect the performance of the anthocyanins. For example there is a certain species of morning-glory which begins its daily cycle in the morning when the cell-sap is slightly acid and the flower colour displayed is a pale pink colour; by the evening, the cell-sap is mildly alkaline and the flower has turned blue. The hydrangea is similarly affected; slightly acid soil causes the flowers to be pinkish whilst under more acid or mildly alkaline soil conditions the flowers turn blue. However, colour variations within species are mainly controlled genetically, and even slight variations in shade may indicate a different genetic composition. The African violet contains an anthocyanin pigment called violanin, which is closely related to the pigment which colours delphiniums. As a result of the thorough research into the genetic behaviour of anthocyanins, African violet growers can now, by careful selection, produce specimens which range throughout the whole spectrum of anthocyanin colouring, from pink to blue, and even including white.

If a petal is examined microscopically, its epidermal cells will often show special peculiarities, especially in the ways in which they are fitted together. Often the cell walls have a wavy and ridged outline, giving the over-all petal surface a beautiful mosaic appearance. The cuticle overlying the epidermis is often striated and very distinct patterns of lines may be formed to give the surface a roughened appearance. Grooves so arranged that they converge towards the centre of the flower, and which may also appear as darker lines, are known as 'honey-guides' since they are thought to direct insects towards the nectaries within the flower. Sometimes these lines may only become visible to our eyes if the flower is photographed using ultraviolet light. This fact leads to the intriguing question of colour perception differences between man's and insect's eyes.

The eye of a bee covers a different part of the colour spectrum from mans, so that bees see flowers in different colours: bees cannot see red, but they can see ultraviolet which man cannot. To a bee, a red poppy appears purplish; other flowers are black but the honey-guides appear as ultraviolet marks signposting the way to the nectaries. Again such flowers as rape, mustard and charlock which all look yellow to man, appear purple, yellow and crimson respectively to a bee, which perceives the little common

Right: This section of a daffodil flower clearly shows the sexual organs—stamens bearing anthers above and below the style, tipped with the stigma and leading to the ovary. See inset for annotation.

Below: Orchids such as this slipper orchid are the most highly adapted of all flowers, with complex mechanisms often linked with a single species of pollinating insect.

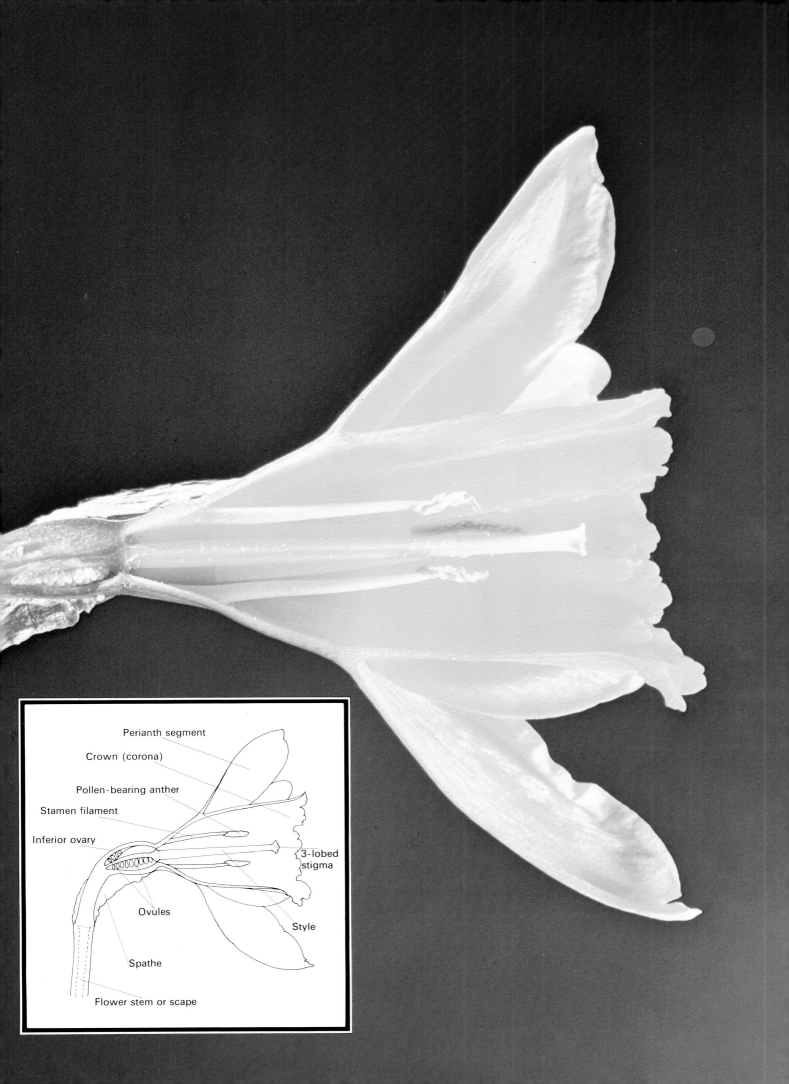

Perianth segment

Crown (corona)

Pollen-bearing anther

Stamen filament

Inferior ovary

3-lobed stigma

Ovules

Style

Spathe

Flower stem or scape

daisy as a halo, since only the tips of its petals radiate ultraviolet.

In an experiment two notices were painted in white letters on a black background, the first being painted with process white which radiates ultraviolet light, the second with Chinese white, which does not radiate ultraviolet. The notices were 'Bees may feed here', and 'No bees allowed here'. Because the second was invisible to the bees, they appeared to be able to read these notices!

Many flowers attract animal pollinators by means of scent as well as by colour. These scents, so familiar in certain flowers, derive from volatile oils produced by certain epidermal cells and their associated hairs, and their range is wide, through sweet, delicate, fiery, fresh, fetid and strong. Often they are a surer way of luring insects into visiting a flower than bright colours are, though most flowers combine the two just to make sure. Scent really comes into its own when the light fails and bright colours start to look drab and inconspicuous. Many of the most intoxicating scents are to be found amongst night-opening flowers, such as the evening primrose (Oenothera odorata), the stocks and the tobacco plant (Nicotiana). Frequently such flowers have expansive white or pale-coloured petals as well. These make the flower appear almost luminescent in the dim light and this, combined with the strong scent, advertise the plant to night-flying pollinators, mainly moths, to best advantage under these poor light conditions.

Not all flower scents are sweet, however. Where flies are the chief pollinators of a plant, smells are often fetid and repulsive to our noses. The arum is one example of this; another is the giant flower *Rafflesia* of Java, the largest flower in the world. This cabbage-like jungle flower, 60cm (24in) in diameter, with its brown and purple colouration, looks and smells like rotting meat, causing hundreds of flies to be crawling over it at any one time. The African stapelias are called carrion flowers because of their odour.

Reproduction

The sepals and petals together constitute the perianth, which is the asexual part of the flower. The sexual part, which functions to produce, bear and protect the sex cells, consists of a male portion, androecium, formed by the stamens, and a female portion, gynoecium, represented by the carpels.

A stamen consists of two parts: a stalk or filament and a swollen part on top of it, the anther, which is usually a bright orange or yellow colour, though sap pigments can produce other colours, as in the tulip where anthers are often purple. An anther has two lobes, each lobe containing a pair of pollen sacs in which the pollen grains are contained. Pollen grains are formed from mother-cells which divide by meiosis (the process by which the number of chromosomes in a nucleus is halved forming a reduced 'haploid' nucleus). Cells packed with starch grains form a layer immediately surrounding the developing pollen, providing it with nutriment, whilst another, outer layer becomes fibrous with lignin-thickened cells. By its uneven contraction upon drying this layer imposes strains upon the pollen sac walls, causing them to rupture. Splitting generally occurs along the grooves in between two pollen sacs where the epidermal cells are very small and easily broken apart. As the anther continues to dry out, so the edges of the now-separate pollen

sacs curl away from the rupture line, finally disclosing their contents of tiny pollen grains to the outside.

When they are mature the pollen grains consist of little rounded cells, each covered in a thick protective wall which is composed of two layers, the exine and the intine. The exine has a thick cuticle and a very irregular surface; it may be covered in spines, pitted or sculptured in a manner characteristic of the species, but not all over, as little pits (germ-pores) are left through which the pollen tube breaks when the grain germinates. The intine, is a much more delicate structure, mainly composed of cellulose and pectic substances, and it is from this layer that the pollen tube itself develops. When it is shed, each little pollen grain contains within itself two nuclei: one is the pollen tube nucleus and the other is that from which two male gametes will arise.

The female organ of the flower is the carpel, essentially a closed hollow container formed by the folding and fusion of a leaf-like structure. Its basic function is to house the ovule which after fertilization becomes a seed. The part surrounding the ovule is the ovary, whilst typically the distal end is drawn out into a prolongation called the style, bearing a sticky pollen-receptive surface at its tip, the stigma. The ovules developing within the ovary are thus completely enclosed by sterile tissue, and it is this feature which essentially distinguishes the angiosperms (the flowering plants) from the gymnosperms (the cycads and conifers).

Ovule development starts with a little outgrowth from a part of the carpel wall called the placenta. The small lump of tissue is called the nucellus and it gradually enlarges and becomes egg-shaped. As it does so, two sets of tissue grow up and around it forming two protective sheaths or integuments. It is from these integuments that the seedcoat is later formed. They do not quite completely enclose the nucellus for a little passage is left at the apex; this opening is called the micropyle. Next the tiny ovule is raised clear of the placenta on a short stalk or funicle which arises from the ovule's base.

Meanwhile, changes have been going on inside the nucellar tissue. A mother-cell formed at the apex of the nucellus divides by meiosis into four, only one developing further, nourished by the nucellus. This megaspore cell develops into the embryo sac, a large vacuolate sac containing eight haploid nuclei. These eight nuclei are arranged in a particular fashion: three positioned near the micropyle end of the sac, the central one being the actual female nucleus or ovum; three more lying at the opposite end of the sac—the antipodal cells; and half-way along the sac are the two remaining nuclei the polar nuclei. The embryo sac, thus complete, brings the ovule into readiness for the fertilization process to take place.

Variations from the generalized flower

There are many variations amongst the flowers to the preceding general description, these variations being useful in the classification of the flowering plants, though they are interesting from other view-points as well. The variations

Right: Special pigments produce the colours of flowers like these in South Africa, which renders them visible at a distance to one kind of insect or another; many groups of insects have particular colour preferences.

cover arrangement, number, size and form of all parts.

The various parts of a flower are arranged on the receptacle in either a cyclic or a spiral fashion, or sometimes a mixture of both. The majority show the cyclic condition, with separate whorls of floral parts. Cactus flowers have their parts spirally arranged, while the buttercup, for example, has its calyx and corolla in two separate cycles and its stamens and carpels spirally inserted. The spiral condition is thought to be the more primitive one. Floral parts usually alternate so that petals do not come opposite sepals but between them, and stamens come between petals and so on.

There is also a difference in the relative positions of parts on the receptacle. This is chiefly related to the different ways in which the receptacle develops. If the final form of the receptacle is a dome, the flower parts are arranged in the order: sepals, petals, stamens, carpels, rising up the dome. The female part of

the flower, the gynoecium, is thus in a superior position relative to the other parts, and this condition is known as 'hypogynous'. The Ranunculaceae (the buttercup family) and Scrophulariaceae (foxglove family) have examples of hypogyny. The earliest flower forms were probably similar to those of the magnolias and the tulip-tree *Liriodendron* which have spirally-inserted parts on a long dome-like receptacle.

Where the receptacle grows into a dish or saucer-shaped structure, the flower parts are as described above only flattened so that all parts are on the same level, with the gynoecium in the middle. This is the perigynous condition and different degrees of it are to be found in the rose family.

The other way in which the receptacle can develop is to become indented into a flask-like shape and in such a case the gynoecium occupies an inferior position relative to the other parts. This epigynous condition occurs in some mem-

bers of the Rosaceae and also in the Compositae (the daisy family) where the receptacular flask is in fact closed by tissues arising from the carpels.

The nectaries, which may be specialized glandular regions or fully-developed glandular outgrowths, may develop as part of the receptacle or on one or more of the floral parts. In the cherry flower, the nectaries are to be found lining the perigynous receptacle, whilst in *Viola,* they are found in the stamen. Buttercup nectaries are tiny outgrowths of the petals.

Most flowers have their parts in definite constant numbers. If there are five sepals, then there are likely to be five petals and five or ten stamens. Five is a common number for the sepals though they do vary from the two-sepalled poppy flower to the water-lily where the sepals are numerous. In the buttercup and rose families, however, the stamens and carpels occur in large numbers ('indefinite') which vary from one plant to another and may even vary from flower to flower.

*Above: A twig of larch (**Larix decidua**) shows the upright pink female and yellow male flowers. The pink is fortuitous, because like all conifers the flowers are wind-pollinated.*

Above left: The inter-dependence of many flowers with insects is brought home to us by fruit-tree blossom like that of apple, for a shortage of bees, for whatever reason, results in very poor crops.

*Left: The carrion flowers (**Stapelia**) are succulents whose blooms produce a disgusting smell of rotting meat to attract the blow-flies which pollinate them.*

In some cases the flower is reduced by one or more of its floral whorls; sometimes the sepals or petals are missing, sometimes even both, and sometimes a flower becomes one-sexed. Petals are entirely absent in the marsh marigold where the five large sepals, instead of being green, are brightly coloured and enlarged, taking on the function of the petals which are therefore no longer required. In the Christmas rose, though the sepals are white and petaloid, the petals remain but are very much reduced to small tubular nectaries. This kind of reduction of parts is an advanced evolutionary character. The other extreme is the multiplication of parts. This is commonly found in many of our cultivated flowers where there is a multiplication of petals and sepals. Although this makes the flower appear more attractive to man, it is a strange fact that many of them are actually sterile.

Primitive flowers have many parts, all separate and freely inserted on the receptacle. This type of condition is characteristic of the buttercup family, the Ranunculaceae. As they advance on the evolutionary scale, flowers generally simplify; sometimes parts go missing as mentioned above, sometimes parts fuse to-

gether laterally, forming petal and stamen tubes as in the cardinal flower. The bird's foot trefoil has nine of its stamens joined together, whilst the tenth is free. Where the number of carpels in a flower is small and definite, these are frequently found fused together as in the gooseberry which has two carpels fused along their longitudinal margins. Some parts, most particularly the stamens, show a tendency to become inserted on an adjacent floral whorl rather than on the receptacle. Flowers with their stamens inserted on the petals are to be found in such families as the Compositae, Scrophulariaceae and Labiatae (the mint family). In some orchids, the stamens are inserted on the carpels instead. Such reduction and fusion of parts makes for greater all-round simplicity in the flower; it is not in any way a degradation for it in fact leads to greater efficiency in the flower, both economically and in pollination processes.

Regular and irregular flowers

A flower may be regular or irregular. A regular flower means that a vertical split through the centre could be made in any one of several planes and the parts would be identical in every case, for example, as in the buttercup. A regular flower is said to be actinomorphic, and is the result of having each part of a whorl identical with all the others and all evenly disposed in the whorl.

The pea flower has five petals. Looking straight towards the inside of the flower, one large and spreading petal stands up as a background to the rest; this is called the standard. Flanking the standard at either side are two wing-like petals, which are therefore called the wings, whilst at the bottom of the flower two smaller petals lie facing each other and together looking something like a ship's keel. They are therefore collectively called the keel. Examining this pea flower, it is clear that there is only one vertical plane through which it could be cut in order to get two halves of the flower symmetrical. The plane would pass vertically downwards through the standard and between the two wings, and bisecting the keel. So the flower is irregular. This irregular condition is also referred to as zygomorphy and is the end result of either fusion of floral parts, absence of one or more parts of a whorl, or differing sizes of parts within a whorl, most often of the petals and/or sepals as we saw in the pea.

Members of the Cruciferae, the cabbage family, show yet another condition. They have flowers which can be divided equally in only two planes at right angles, and the half-flowers produced by cutting in one such plane are different from those produced by cutting in the other, though the flower is still considered to be actinomorphic. Flowers with spirally inserted parts, such as in the cacti, cannot be divided into two exactly equal halves however many cuts are tried, and are hence termed asymmetric.

Flowers with irregularly shaped petals are often the result of specialized adaptation to a particular pollination method; there is division of labour within the petal whorl. The upright standard and flanking wings of the pea for example serve together to advertize the flower whilst the keel functions as an attractive solid-looking landing base for a visiting insect, the weight of which upon alighting depresses the keel downwards, exposing the sexual parts which stick up onto the underside of the insect.

The flowers of the family Compositae, which includes daisies, dandelions, hawkweeds, deserve special mention. What at first appears to be a petal in the flower head is actually a complete flower, often called a floret. These florets are all tightly packed together to amass a broad expanse of colour to attract visiting insects. In some Compositae there is a sharp division of labour amongst the florets. For instance, in the black-eyed susan, the outer insect-attracting yellow ray is composed of sterile flowers with conspicuous petals, whilst the inner circle is made up of hundreds of fertile ones, with/their tiny inconspicuous fused petals, ensuring efficient pollination.

The ways in which flowers are arranged on a plant is termed the inflorescence. Flowers may develop singly or in groups. They may be at the end of the main shoot or one of its branches, with or without a stalk. Amongst insect-pollinated flowers it is common to find small flowers grouped together to provide a visually more attractive expansive inflorescence, for example the much-stalked, flat-headed type of inflorescence characteristic of the carrot family, the Umbelliferae. Single flowers tend to be larger and more showy, such as the tulip or the wood anemone.

Most flowers are hermaphrodite, that is they possess both male and female parts, the stamens and the carpels, together in the one flower. However, some flowers possess either stamens or carpels, but not both, and are thus unisexual. Male and female flowers may occur together in the same inflorescence, or there may be unisexual flowers together with 'normal' hermaphrodite flowers in the same inflorescence; both

*Above: The 30cm (12in) flowers of **Protea cynaroides** are composed of numerous florets surrounded by bracts, brightly coloured to attract the birds that pollinate them and suck their nectar.*

*Above left: The sexual parts of the moth-pollinated **Hibiscus rosa-sinensis** are formed into a long tube in which the stamens surround the projecting style, ensuring that pollen reaches the moth.*

*Left: The huge blooms of the Indian sacred lotus, **Nelumbium**, are carried well above water level. This widespread water lily relation is pollinated by beetles and similar insects.*

these conditions are to be found in the Compositae and the Gramineae (the grass family).

Sometimes separate male and female flowers occur together on the same plant but in different inflorescences. This happens in the hazel and oak trees, and such a species is called monoecious. Dioecious species, which include willow and poplar, have male and female flowers on separate plants. Occasionally, as in the Compositae, flowers develop which have neither stamens nor carpels. These flowers are thus neuters, which may seem a contradiction of terms, in view of the picture drawn of flowers as reproductive devices. However it must be remembered that such flowers are in fact highly specialized for their purpose in life; they direct all their energies into being attractive, thus luring insects to the fertile flowers, with which they are in close association.

POLLINATION
the act of reproduction

Pollination is the process whereby pollen grains are transferred from the anthers of a stamen, their site of origin, to the receptive part of a carpel, ie the sticky surface of a stigma. The male sex cells, which are contained in the pollen grains, are still separated from the female sex cells in the ovules, enclosed in the ovary, by the style. Thus further processes are necessary before fertilization can occur. Funda-mentally, pollination is merely a preliminary to fertilization, but a very necessary preliminary.

Cross-pollination versus self-pollination

Pollen may pass from the anther of a flower to the stigma of the same flower or another flower on the same plant, resulting in self-pollination; or it can be transported to the stigma of a flower on another plant of the same species—cross-pollination.

From the many and varied devices, some amazingly elaborate, which plants have adopted to secure cross-pollination, it can be assumed that for some species it is advantageous, and for some obligatory since they will not set seed with their own pollen. Seed produced from the union of two different parents will combine some of the individual features of both, thus bringing about a greater possibility of variation in the offspring. This variation is necessary because without it there can be no further adaptation to a different or changing environment. Stronger or better-adapted individuals will tend to survive, while the weaker ones will be eliminated in the course of competition for survival.

However, too much variation can create instability which for the species as a whole can mean a failure of adaptation to either a relatively stable environment or one that is changing slowly, in addition to a possible loss of a vigorous and well-balanced constitution already achieved. This is probably why there are still today a large number of plants which are able to resort to self-pollination should cross-pollination fail to take place. This can be achieved with the aid of structural and be-havioural devices favouring cross-pollination in the early stages of flower opening, followed by the possibility of self-pollination if cross-pollination has failed. For example, love-in-a-mist, *Nigella,* is usually cross-pollinated. The styles of its carpels are much longer than the stamens, so that it is impossible for the ripe pollen to fall on the stigmas and bring about self-pollination. But if, when the pollen is ripe, cross-pollination has not occurred, the styles gradually bend backwards until they come into contact with the exposed pollen on the anthers so effecting self-pollination.

Before the variety of devices which plants have evolved through the ages in order to further the possibility of cross-pollination and obstruct self-pollination are examined, how pollination is actually effected (that is what agencies are involved) should be discussed. Ovules are fixed and stationary inside ovaries. Pollen grains when mature are exposed freely on the anthers either as separate little dust-like particles or stuck together in small groups, but they are not themselves able to move. At its simplest, pollination takes the form of pollen grains falling by the gravitational pull from an anther onto a stigma, or of the stigma and anther drawn into contact with each other by their manner of growth or by alteration of relative positions at maturity. More frequently, outside agencies are needed to effect transfer. To further this end, plants have made use of almost every thing that moves in their environ-

Prunus dulcis
Almond

Right: Attracted to flowers both by nectar and pollen for food, the bee is one of the important pollinators; but insects of every size from midges to giant moths perform this function.

ment, both animate and inanimate. Thus these agencies include a variety of 'handy' animals, including birds, bats, slugs and snails as well as insects, the wind, and even water. Of these agencies, insects and the wind tend to predominate, especially in temperate climates.

Bird-pollination

Bird-pollinated flowers are common in warmer countries, the bird species most particularly involved being the humming birds, honey-eaters, honey-creepers and certain kinds of parrots. A flower adapted to bird-pollination is often apparent by its vivid red or orange colouration and by the fact that it contains a very plentiful supply of nectar within it in order to make the visit worthwhile from the bird's point of view. The habitually bird-pollinated American flower, *Erythrina cristagalli,* is known as the cry-baby flower because it contains so much nectar that it drips out of its inflorescence, while species of *Banksia* in Australia produce so much nectar that it is used as food by the aboriginals. Reds and oranges prevail because birds have a preference for harsh colours; some bird-flowers have rather peculiar colour combinations ('parrot colouration') such as green, yellow and scarlet mixed together, a combination frequently found in the pineapple family. No scents are produced because birds have very little sense of smell.

Although humming birds can be very tiny and some bird-flowers can be extremely large, birds usually are larger than the flowers they pollinate and tend to be rather clumsy in their dealings with them. To counteract this, many bird-pollinated flowers have stamens strengthened with woody filaments to minimize damage and the ovaries are frequently protected from accidental pecks either by being inferior (that is below the corolla, anthers and stigmas) or by being separated from the nectar-secreting part. Sometimes, as in *Fuchsia,* the flowers have flexible stalks so that they get pushed away if the bird probes too forcibly, and thus they escape damage. Often the plants provide portions of bare stem, leaf stalks or some such place for birds to alight upon, though some Australian plants, for example *Brachysema,* are pollinated by birds standing on the ground. American bird-flowers tend to have longer flower tubes than Asiatic ones because American pollinating birds tend to have longer bills. Occasionally bird-flowers are pollinated by rats, squirrels or, in Australia, by small arboreal marsupials, too.

Bat-pollination

Pollination by bats is quite important in the tropics. Because bats tend to be crepuscular in their activities, we find that the flowers which they pollinate are night-opening. Such a flower is *Crescentia cujete,* the calabash tree. Bat-flowers often have dingy colours and a sour musty scent (rather a 'cabbagey smell'). Although bat-flowers have evolved independently in the Old and the New World, the scents of both groups are similar; the smell being similar to the peculiar odour of the glandular secretion which most bats produce for recognition purposes. Bat-flowers evolved by imitating this smell. Generally bat-flowers produce a lot of sticky nectar, though sometimes fleshy sugary bracts are provided for food as in the Indonesian plant, *Freycinetia insignis,* which is pollinated by fruit bats. Bats visiting flowers often cause damage with their claws: such flowers tend to be large and strong to help guard against this.

Slugs and snails as pollinators

Creeping animals such as slugs and snails unwittingly pollinate some flowers by picking up pollen grains in their slime and depositing them on stigmas encountered along their way as they move over plants. Some flowers are in fact specifically designed for pollination by slugs and snails; *Rhodea japonica* of the lily family has fawn-coloured flowers smelling of bad bread which provide fleshy tissues for their visitors. The aspidistra, with soil-level flowers, is also pollinated by snails.

Insect-pollinators

Other animals may inadvertently help disperse pollen grains too, but long ago plants discovered that it was the insects which were to become their most reliable allies, particularly the bees, wasps, butterflies, moths, flies and beetles. These are useful pollinators because with their small size they are able to alight upon and enter flowers without damaging them, and they can easily collect pollen on their hairy backs and legs when visiting the flower for food. As the insect passes from flower to flower, the pollen picked up from the stamens of the first is passed onto the stigma of the next.

Insects are attracted initially to the flowers by the bright colours and delicate perfumes of the petals and/or sepals, and come to associate these with the presence of nectar for their taking. Insects appear to have developed their acute sense of smell in close association with the evolution of the flowering plants, since the insect ancestors 250 million years ago, long before the flowering plants had appeared, had little sense of smell, having no need for it then. Many insect-pollinators have an acute sense of taste also; the butterfly for instance contains its organs of taste in its feet and if a single hair on one of the feet touches a nectar solution, it is sufficient to bring the long tubular tongue into action. Many plants show maximum nectar-secretion around midday to coincide with the peak of insect activity at that time, while many bees will organize their flower-visiting routines to coincide with flower-opening times.

Some insects visit flowers not for nectar but for the pollen which is their food. Flowers which are visited in this way include various species of poppy *(Papaver),* and species of *Helianthemum,* the rock-roses. Pollen is often found in the honey stored by insects—hive-bee honey owes its vitamin content to the pollen which it contains. In some beetle-pollinated flowers, food is provided in the form of special tissues which are filled with starch and oil, usually attached to the stamens. Sometimes the beetles can stay there for days, eating away.

A simple open flower with spreading petals, such as may be found in the buttercup family, the Ranunculaceae, attracts and provides a landing stage for a wide variety of insects, both large and small; butterflies, moths, bees, wasps, bugs, beetles and flies are all to be found on buttercup flowers. So the buttercup is not specialized in any particular way towards a selected type of insect pollinator. Many flowers however are precisely adapted to pollination by only one kind (or by very few kinds) of insect and some flowers evolved a tube-like form to conceal the nectar. This can only be reached by long-tongued insects with good three-dimen-

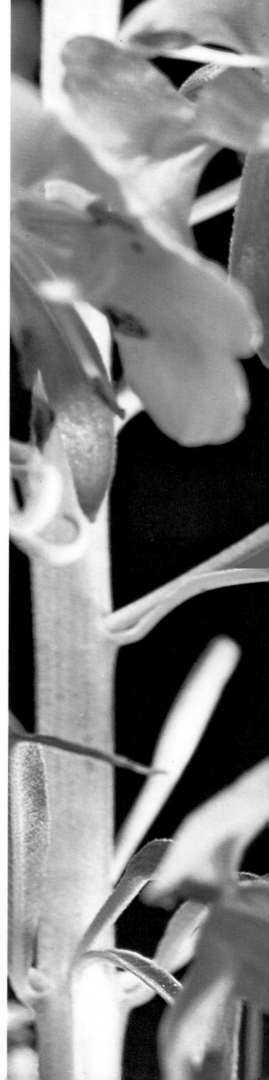

Right: The willowherb (Epilobium) shows an example of pollination 'insurance'. If the flower fails to be pollinated by insects the styles curl back to self-pollinate.

Above: Sections of bogbean (Menyanthes) flowers show long stamens and short style at left, and the opposite at right—a mechanism designed to favour cross-pollination.

Left: A single birch catkin produces about five and a half million pollen grains, which are dispersed by wind. It is a tree of northern latitudes where pollinating insects are scarce.

sional perception, such as bees which are more efficient pollinators than short-tongued beetles and others. The periwinkles (*Vinca* species) have the corolla in the form of a slender tube, with a flat part on which the visiting insects can alight, and many flowers in the olive family (Oleaceae) are similarly constructed.

Other flowers have become even more specific in their pollinators. Strange cactus-like plants (*Stapelia* species) of Africa and southern Asia, with large hairy dark purplish-red flowers which look and smell like bad meat, attract only flies. The Old World orchid, *Bulbophyllum macranthum,* smells of cloves and attracts only a single species of fly. In Southeast Asia, each species of fig-tree is pollinated by its own particular species of fig-wasp. The bee-orchids (*Ophrys* species) have a petal lip which is shaped, coloured, sized and even smells the same as the body of the female of certain kinds of humble-bee and related insects. These flowers attract the male insects which try to mate with the orchid lip. The pollen in orchid flowers is clumped together in a pair of club-shaped pollinia (the plant in effect 'puts all its eggs in one basket') and these stick to the head of the

humble-bee during its activities and get carried away to pollinate another bee-orchid.

The monkshood (*Aconitum*), too, is dependent on humble-bees, for only these can reach the nectar underneath the plant's 'hood' and in so doing effect pollination. *A. lycoctonum* of the Alps in fact has nectaries some 20mm (0·8in) long which can be exploited only by a few of the very longest-tongued humble-bees. The monkshood has become so dependent on humble-bees that it is not found outside the insect's range. This is the disadvantage of specialized adaptations towards one kind of insect pollinator; although it makes for greater precision in the pollination act, the plant becomes so dependent on this insect that pollination will suffer should the local insect population be adversely affected in any way by factors beyond the control of the plant, such as bad weather, disease or insecticides.

A curious relationship exists between pollinator and flower in the case of the American yucca plant. This flower is pollinated by a single species of moth *Tegeticula yuccasella* which is attracted by the large creamy white night-scented flowers of the yucca. The female moth climbs up a stamen, bends her head closely over the top of the anther, steadies herself with its uncoiled tongue, and then scrapes all the pollen off the anther, into a big lump which she carries under her head. She may do this to as many as four stamens. The yucca-moth then flies to another flower and investigates the condition of the ovary carefully. If she considers the flower suitable, the moth again climbs up the stamens, but this time goes between them to the ovary into which she bores a hole and lays an egg. Immediately after doing this, she climbs

up the stigmas, which are united in a tube, and thrusts some of the parcel of pollen down the tube. This behaviour is repeated with each egg that is laid. This procedure ensures that there will be food, provided by abnormal ovule growth next to each moth's egg, for the larvae; an unpollinated yucca flower soon dies. In this relationship, the moth ensures the seed-production of the food-plant, while the plant provides food and shelter for the young of its pollinator.

Wind-pollination

Another important agency for pollination is the wind. Plants specially adapted to this method include many trees, grasses and grass-like species, and many sea-side plants. Conifers come into this category, their clouds of pollen blowing in the wind like yellow smoke, being a common sight in early summer. This method has the advantage of freedom from the unpredictable activities of insects; it is effective when insects are scarce or absent. However the chances of a pollen grain reaching an appropriate stigma by wind are so remote that many thousands of times more pollen grains must be produced than can actually be used for fertilization. It has been estimated that a single birch catkin produces about five-and-a-half million pollen grains and a single floret of rye over fifty thousand grains. An incidental effect of all this pollen released into the air is that it may set up allergies in some people causing the common condition of 'hay-fever'.

Insect-pollinated species tend to have rather sticky and highly sculptured pollen grains, to attach themselves to insects bodies, whereas in wind-pollinated plants the grains have smooth dry surfaces and are dispersed singly or in two's or three's. Pollen grains so constructed can be carried great distances; they have been found in the air even in mid-Atlantic!

Wind-pollinated flowers share several features. They often have large stamens which hang out of the flower on long filaments or in catkins so that they are freely exposed to the air. Stigmas too are generally large and well exserted; and they may be finely feathered so that they can 'net in' pollen from moving air. Petals and sepals are inconspicuous often reduced or absent altogether since they cannot 'attract' the wind; in fact they are a hindrance by obstructing the air flow and free transfer of the pollen. Often the sexes are separated in different flowers to minimize the chances of self-pollination. Many wind-pollinated deciduous forest trees flower very early in the year. At this time, the trees are bare of leaves so that the movement of pollen grains meets with least obstruction. At this time too, very few insects are active; perhaps in temperate regions, effective pollination of these dominant plants would require a greater population of insects than such a climate could support.

Water-pollination

Pollination by water is rather rare. Most water plants have their flowers above the surface where they can be insect-pollinated. Species that are water-pollinated mostly show a relatively recent ancestry of insect- or wind-pollination. As with wind-pollinated plants, water-pollinated flowers tend to have reduced petals and sepals, a single ovule and flowers which are commonly unisexual. Stigmas tend to be large, but rigid and simple, whilst pollen grains are often elongated; it is interesting that

some aquatic fungi (the Hyphomycetes) also show a trend towards elongation in their spores.

Water-pollination may take place at the water surface or completely submerged. A good example of the former is the ribbon-weed, *Vallisneria spiralis,* with male flowers which break free and float to the water surface where they open and the stamens release their globular masses of pollen. The separate female flowers are borne to the surface on slender flexible stalks where they open in a little depression of the surface-film. They disclose three large fleshy stigmas which are unwettable due to a covering of water-repellent hairs forming a dense velvety pile. The projecting stamens of the male flowers are brought into contact with the stigmas of the female flowers at the water surface by water currents, the male flower sliding down the little depression in the surface-film.

In the eel-grasses *(Zostera),* the pollen 'grains' are thread-like, about 0·25mm (0·01in) long and have the same density as sea water so that they may move freely at any depth until caught on the submerged feathery stigmas of the female flowers, round which they rapidly curl.

Mechanisms to ensure cross-pollination

Cross-pollination is a more effective method of producing new variants than self-pollination. Plant breeders can produce vigorous 'hybrids' by using a little brush to transfer pollen in a controlled method of cross-pollination, but in nature many diverse mechanisms exist to favour cross-pollination and obstruct self-pollination.

Self-sterility is not uncommon among flower-ing plants; if pollen reaches the stigma of the same flower, it is unable to develop, or can only do so very slowly, because the stigma inhibits its growth. Buttercups are markedly self-sterile. This 'prepotency of pollen from a distinct individual over that of the same individual' was discovered by Darwin in 1876.

Lots of plants have their anthers maturing long before the stigmas of the same flower are sufficiently developed to accept pollen, so that pollen from the flower has already gone before it has the chance of pollinating its own stigma. These flowers are called protandrous, for example the dandelion *(Taraxacum)* and daisy *(Bellis).* Some flowers have stigmas which are mature before the stamens, but these are less common, though the mechanism can be found in the meadow rue *(Thalictrum flavum)* and the woodrush *(Luzula),* and these are referred to as protogynous flowers. This condition in which anther maturity and stigma receptiveness do not occur together is known as dichogamy.

Cross-pollination is the only possibility if individuals of a species produce gametes of one kind only. Most animals are unisexual, and some plants are too, having separate male and female flowers either on the same plant or on separate ones. Many wind-pollinated flowers are unisexual since the stamens here produce such vast quantities of pollen that if stigmas were close by, they would be so thickly covered with their own pollen that there would be little chance of fertilization by pollen from other individuals.

Heterostyly is another device by which some plants seek to avoid self-pollination and the primrose *Primula vulgaris* is a fine example of this. Primrose flowers from different plants are

Above: Enormously magnified with a scanning electron microscope, the individual pollen grains are seen to be covered with spines. It is thought that these spines aid the grains to stick to insects which transport the pollen from one flower to another, and make them adhere to the receptive hairs of the females stigma. A fractured pollen grain (above right) shows the germination pores through which the pollen tubes emerge, to travel down into the female ovary.

Right: The stamens on mallows like Lavatera are united into a long tube around the projecting styles. However, when visiting bees alight upon this organ they are dusted with pollen.

of two kinds; those with the top of the stigma protruding like a little pin-head up through the throat of the corolla tube and those with five anthers clustered around the neck of the tube and the stigma invisible beneath them. The former flowers are called pin-eyed, the latter thrum-eyed, and the difference is essentially due to the style being long and the anther filaments short in the pin-eyed flower and vice versa in the thrum-eyed. The primrose is pollinated by the honey-bee which when pushing its long tongue into the corolla tube of a thrum-eyed flower will receive pollen in the right position for pollinating the stigma of a pin-eyed flower, and vice versa if it visits a pin-eyed flower. Also the pollen grains of the pin-eyed flower are smaller than those of the thrum-eyed, and its stigma has coarser surface papillae. The larger pollen grains of the thrum-eyed flower tend to stick more readily to these

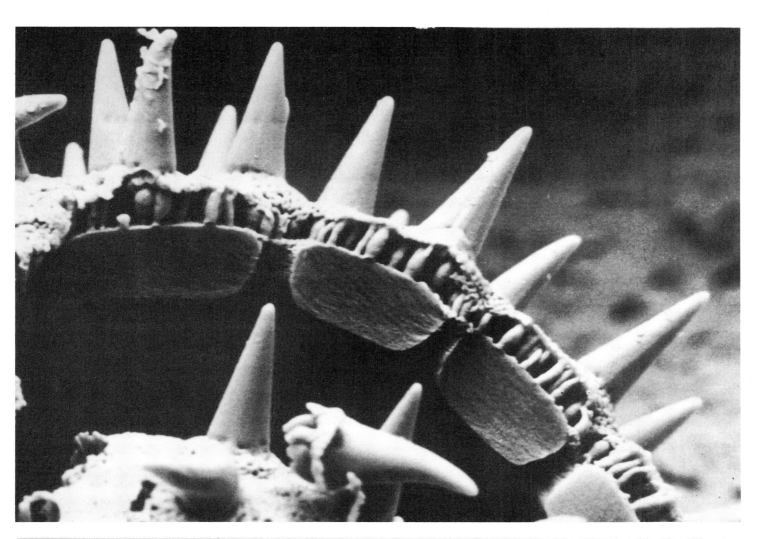

coarser papillae, whilst the smaller pin-eyed pollen grains are better suited to the thrum-eyed stigma and its finer papillae.

Some insect-pollinated flowers have devices which detain the visiting insect longer, so that pollination can be more effective. Classic examples of this are to be found in the bind-weeds and trumpet gentians which have been called 'revolver flowers' because upon entering them, the insect is faced with a ring of narrow tubes like the barrels of a revolver, through which the insect must probe to get its nectar. Other flowers actually trap their insects. This happens in many members of the birthwort family (Aristolochiaceae) and in the Araceae to which *Arum* belongs. Essentially the mechanisms are the same in that lured insects slip down the smooth-sided flower tube which has smooth-surfaced cells and downward-directed papillae coated with a layer of oil, and often downward-pointing hairs too, all of which prevent upward escape. A day or two later changes take place which enable the insect's liberation—the trap hairs and papillae shrivel and the tube may

The eyes of bees are sensitive to ultra-violet light, and what the bee perceives is not always what we see. To us flowers of evening primrose (Oenothera) appear yellow (right); but to the bee they appear blue, as this ultra-violet photograph shows (below). Note also how the faint gold flush in the centre of the flower is in fact blackish to the bee; this black centred 'star' in fact helps the insect to find the sex organs and nectar in the centre.

Wild arum-the tender trap

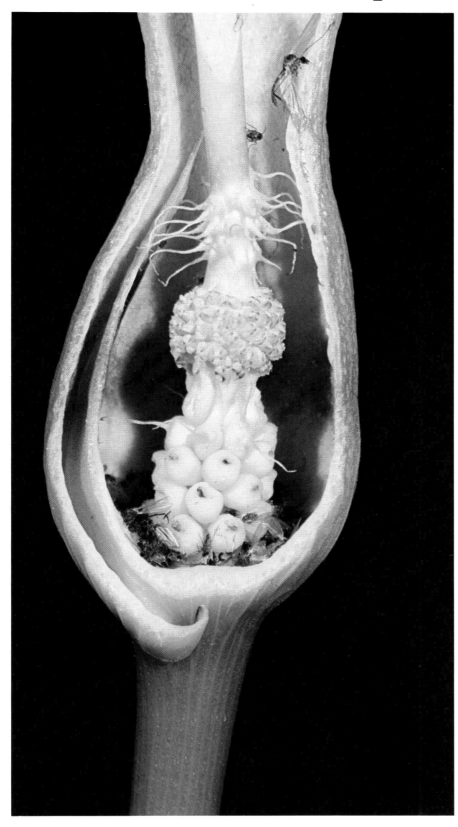

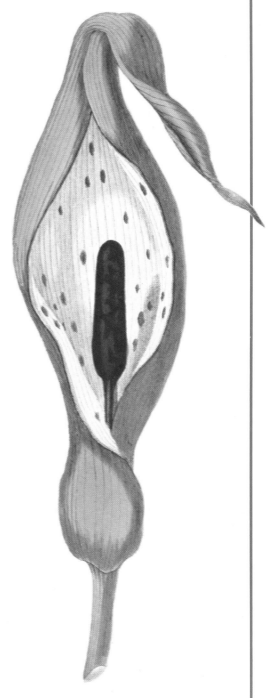

The temperate Wild Arum or Cuckoo Pint **Arum maculatum** bears a curious inflorescence which attracts and traps flies in order to ensure pollination. The plant produces flowers in the spring which consist of a central 'spadix' and an enclosing leafy sheath or 'spathe'. The spadix bears simple female flowers at the base and a zone of male flowers further up, separated by a whorl of sterile flowers. A second zone of sterile flowers with long bristles are above the male flowers and guard the lower chamber which is formed by the enclosing spathe. The upper part of the spadix is club shaped and develops an unpleasant odour when the upper portion of the spathe unfurls. The scent attracts flies and small beetles that fly into the flower but are unable to keep their feet on the smooth surfaces of the spathe and spadix. Small insects fall through into the flower chamber while larger flies are arrested by the guarding bristles. At first the female flowers are mature and receptive to any pollen carried by the flies. Later the anthers of the male flowers break open and shower pollen on the captive insects. Finally the flower's guard bristles wither and release the flies, hopefully to visit other plants and get trapped again.

change to a horizontal position as in the midge-trapping *Ceropegia,* a genus occurring in Africa, Asia and Australia. In the South American birthwort *Aristolochia lindneri,* flies are trapped on the first day of the flower's opening, on the second day no luring (to the fly) smell is produced and the stigmas bend together so they can no longer receive pollen, and just before the insects are released the anthers burst and dust the prisoners with pollen. In *Arum,* pollen deposited on the stigmas on the first day produce rapidly-growing pollen-tubes so that the stigmas are withered by the time the inflorescence sheds its own pollen. By these means self-pollination is prevented.

The insect-pollinated flowers as a group in fact show some truly amazing mechanisms for securing cross-pollination. One such flower, an orchid called *Gongora maculata,* has a very slippery corolla, plus an intoxicating (to a bee) scent which combine to make a visiting bee lose its feet and slide down a curved and downward-hanging structure called a column. As the bee slides down the column it picks up on its way the pollinia pollen-clumps conveniently hanging in its path. Should the bee now visit a female flower, the bee is exposed to a sticky stigma which rubs the pollinia off its body as it slips down the slide.

Other orchids—a group which is highly specialized for insect pollination—show some equally fantastic pollination methods. The massive waxy-looking flowers of *Coryanthes macrantha* have part of the corolla lip forming a 'bucket' into which fall drops of water secreted by a pair of knobs on the column. Male bees of one particular genus *(Euclema)* are attracted to this flower by its strong scent

which leads them to a special place on the flower lip where they endeavour to collect the liquid scent by scratching at it with their forelegs. This fluid which they collect affects special sense organs on the bee's feet, causing intoxication, so that it loses its grip and falls into the 'bucket' of water in which it swims around, but it cannot climb out again. The bee has one exit, however, through a narrow tunnel just beneath the stamens and stigma. As it wanders drunkenly out through the passage, the bee picks up or deposits pollinia. By the time the bee has escaped, the flower's scent has vanished, so that the bee cannot be induced to re-enter the flower, though the scent returns the following day. By this prevention of re-entry of the same flower, self-pollination is avoided.

One group of plants are, however, habitually self-pollinating. These are the annuals, small plants often living in unstable habitats, which also tend to be rather restricted and precisely defined. Self-pollination in this case is advantageous in that it maintains their precise adaptation to demanding habitats of this type. Also because their habitats are unstable and large proportions of annuals can be wiped out in a bad year, self-pollination helps numbers to recover more readily and quickly. Cross-pollination may occasionally occur, helping to provide some variants.

Some plant species make the best of both kinds of pollination in the same individual plants by producing two different kinds of flowers. The violets *(Viola* species) have conspicuous coloured flowers which are in every way adapted to cross-pollination by insect visitors, also possessing devices to ob-

struct self-pollination. Seed-setting is however erratic and often poor. If the plants are examined following flowering, little things which appear to be flower-buds on short stalks can be found. These are actually flowers in their own right; they remain small, never open and are self-pollinating, with seeds being set in abundance.

Right: In warm countries many flowers are pollinated by birds, like this lantana being approached by a humming bird. Such flowers usually have copious nectar which the birds suck up.

Below: Many tropical trees are pollinated by bats like this Long-nosed Bat from central and south America. It has a long, brush-tipped tongue for feeding on pollen, and correspondingly reduced teeth.

Bottom: Flowers are usually highly adapted to their pollinators, and these visited by butterflies with long slender prosbosces always have very narrow tubes into which the insect pushes its long 'tongue'.

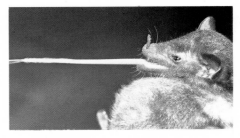

SEEDS
the ongoing generation

Seeds represent the culmination of the re-productive cycle of flowering plants. Many flowerless plants produce them as well, for they are a very efficient type of dispersal structure. They existed over 250 million years ago and some of the earliest seed bearers still flourish today, such as the conifers and cycads (or 'sago-palms').

Not all plants, however, produce seeds, some rely on spores for reproduction and dispersal.

Fertilization

Pollination is an essential preliminary to the union of the male and female gametes in the fertilization process. When the pollen grain containing its male gamete is resting on the stigma surface, still separated from the ovule and its female gamete by the length of the style and the carpel wall, the pollen grain must somehow 'bridge the gap'.

But before the pollen grain can show any activity, certain conditions have to be met with. Firstly, nothing will happen unless it is on the stigma of its own species, though exceptions occur, resulting in rare hybrids such as the cabbage–radish cross. For normal pollen development, a sugar solution of specific concentration, particular to each species, is necessary. This is somewhere between 5 and 30 percent cane sugar.

In favourable conditions, the pollen grain pushes out a delicate tube, which is in fact a bit of the intine which appears through a hole in the exine. Energy for this is provided by the rich store of protein and fatty substances in the pollen grain. The pollen-tube has a natural tendency to grow away from the light and air, so it disappears into the stigma, and starts to wend its way through the stigmatic tissues, partly by pushing a path between the cells, which are usually loosely arranged, and partly by dissolving the cells by means of enzymes. The method which predominates depends on the species. The digestion of cells helps provide extra energy for the developing tube, which is guided on its long journey by a chemical substance of a sugary nature given off by the ovule. As it travels down the style, the tube carries the pollen-tube nucleus like a pilot at its tip with the two male gametes right behind. Eventually the carpel cavity containing the ovules is reached. Now the tube tip penetrates an ovule, usually via the micropyle, opens and discharges its contents into the embryo sac. The tube nucleus disintegrates, its business over, one male nucleus fuses with the female nucleus, or ovum, thus initiating the embryo, and the other fuses with the two polar nuclei, bringing the 'endosperm' into being. In conifers, the ovum alone is fertilized.

This fertilization process in plants is thus both a complicated and a remarkable feat. Considering the pollen-tube development alone, it is quite astonishing to compare the small size of the pollen grain with the length of tube it produces; in corn, the pollen tube has to grow through the silk and whole length of the ear in order to reach the embryo sac, a distance of some 30cm (12in), while the pollen grain itself is little more than one-four-hundredth of a centimetre.

Development of seeds

Once fertilized, an ovule starts to develop into a seed. It enlarges, and the embryo, by repeated cell division, forms a flat structure called a cotyledon. A seed produces either one or two of these cotyledons. In the former case the plants are called monocotyledons, a group which evolved later than the second group, the dicotyledons. This division amongst the flowering plants is rather more apparent later on; monocots most often have grass-like leaves with parallel veins, whilst, typically, dicot leaves are broad with herringbone or finger-like veins. There is a difference in the number of flower parts also, monocots mostly having them arranged in threes, whilst fours and fives, and infinite numbers, are common in dicots.

Back in the seed, the developing embryo is pushed into the seed centre by a sort of tiny elongating pedestal. Here it becomes surrounded by the endosperm which provides nourishment during its development. Many dicots absorb the whole of the endosperm, and the cotyledons end up thick and fleshy, taking up almost all of the seed, as in the broad bean and

Ecballium elaterium
Squirting Cucumber

Right: One of the biggest seeds, the coconut can float and is washed up on sandy sea-shores. Its food reserve lasts for up to two years so that it can usually establish itself.

the pea. In most monocots, and some dicots, the bulk of the reserve food rests in the endosperm until the seed germinates, as for example in maize, onion and date seeds. The seed now has its embryonic seedling leaves in the form of its cotyledons, an embryonic shoot called the plumule and an embryonic root called the radicle. In most seeds the embryo itself is very small—less than one millimetre long in fact. The seed has become invested in a protective coat or testa, formed from the old integument of the ovule and, with its essential structure formed, the whole seed undergoes drying-out so that its water content may drop to a value as low as 10 percent by weight of the seed. Finally

Right: After germination the peanut seed opens into two halves, the young plant emerging between the thick food-packed cotyledons which become green on their inner surface.

Far right, above: Cress seeds germinate very rapidly in warm moist conditions, as growers of 'mustard and cress' know. The first organ to emerge is the root; the green seed leaves follow.

Far right, below: This young beech seedling shows the paired cotyledons or seedling leaves and the first pair of adult leaves, which are quite different in shape.

Below: Mangrove seedlings germinating at the edge of a salt-water lagoon. They start sprouting while still on the parent and are designed to drop straight into the mud.

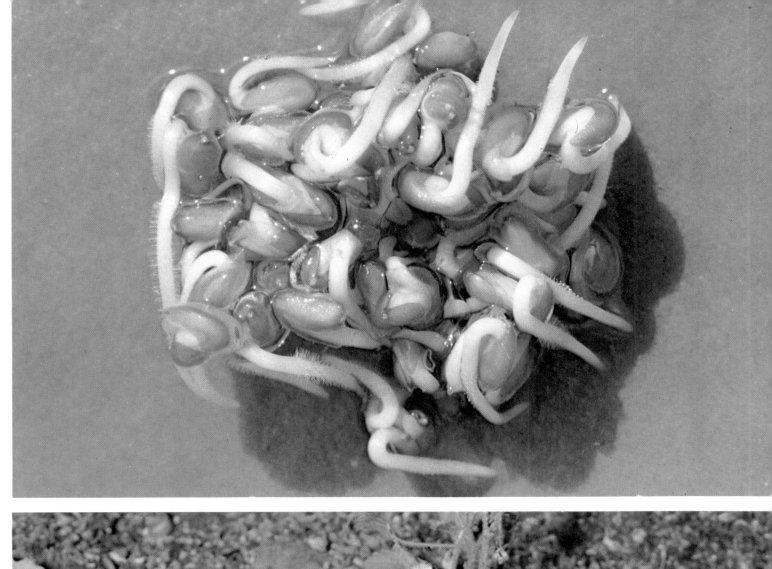

the mature seed passes into a dormant condition. Growth stops, respiration slowing down to a point where it is no longer detectable; the seed in fact passes into a sort of suspended animation.

A shelled peanut reveals seed structure nicely. Around the outside lies the protective brown testa. Inside, the two halves of the peanut are the cotyledons, fitting snugly together. At one end the cotyledons are hinged together at the point of their attachment to the embryo, which lies tiny and perfectly sculptured between the two halves. Upon germination, this small embryo will start using the food stored in its cotyledons, and begin its development into a whole new peanut plant.

The seed is a complex structure. Even the tiniest seed consists of many cells; larger ones already contain a highly developed embryo before they leave the parent plant. All contain a food store, either in fleshy cotyledons or the endosperm. The supply may be quite substantial, hence the food value of such seeds as peas, beans and nut kernels to man. The testa protects its contents from excessive heat and light and prevents water-loss. Clearly, then, the seed is well prepared for its journey away from the parent plant. A spore, by comparison, is simply a single cell with a relatively flimsy cell wall. It contains only very limited food resources, and its chances of surviving and growing into a new plant are correspondingly very small. Consequently, a spore-producing plant must devote a considerable proportion of its energies into producing sufficient of these particles to compensate for the all-too-likely losses.

Conditions necessary for germination

In its dormant state the seed can withstand extremes of temperature, way below freezing or near the boiling point of water. How long the seed remains in this condition depends on the particular plant species. In the tropical mangroves, there is no dormant period. Here germination takes place whilst the seed is still held within the fruit, and the young seedling

falls like a dart with its long pointed radicle straight into the mud of the mangrove swamp. Other seeds actually need a long dormant period before they are able to germinate. Certain other conditions must be fulfilled, too, for germination to take place: water and oxygen must be readily available and the temperature must be within a certain range, depending on the species but usually between 5 and 30°C (41–86°F). Some seeds require other, special conditions, such as the water plantain *Alisma plantago*, the testa of which has to decay or become damaged before the embryo can emerge.

Under very dry conditions, seeds can remain viable for quite a time, but there has been a tendency in the past to overestimate their longevity. Under normal conditions, seeds can be stored for only one or two years. After this time the natural humidity of the air makes them dissipate their store of energy and they can no longer germinate. A closer look at the triggers of seed germination reveals why this is so.

The seed begins to 'stir' when water penetrates the outer seed coat and the water content of the seed comes to reach around 8 percent of its total bulk. This is sufficient to support respiration but little else. Below this level the seed remains inert. Above 12 percent there is sufficient water for both active life and the initiation of growing. In ordinary air, there is usually sufficient moisture for very slow respiration, but not enough to support growth. Without further water, the seed will slowly burn up its food reserves, until its potential for

Right: One reason the spear thistle, **Cirsium vulgare,** *is such an efficient weed is its profusely produced seeds which are dispersed on the wind with the aid of a feathery pappus.*

Below: A close-up of the large pappus below which the seeds of the goatsbeard (**Tragopogon** *) are transported on the wind. Small feathery hairs are interwoven on the main 'umbrella'.*

Above: The woody pods of laburnum twist apart violently when quite dry, making a sharp report and throwing out the large seeds which can be projected many feet.

Left: The 'keys' of sycamore seeds are beautifully designed as aerofoils: once ripe and separated, they spin slowly down to ground level, and in a wind may land some way from the parent.

germination runs out, and the seed dies. Seeds with firm hard coats tend to retain their viability for longer. The best conditions for the storage of most seeds are low temperatures and a dry atmosphere; in this way, seeds of common farm and garden plants have kept for 10 to 25 years.

A few years ago, stories were circulating concerning wheat seeds taken from the desert tombs of ancient Egyptian pharaohs which grew upon being sown, 4,000 years after their formation. Sadly, the stories were untrue. Even in the ancient sarcophagi some moisture exists, and wheat seeds last only a short time before internal changes render them incapable of germination. A well-authenticated case of seed longevity concerns some seeds of the Indian lotus (Nelumbo nucifera) which had been kept in the herbarium of the British Museum. In 1940, the museum was bombed, and the seeds were soaked with water which resulted in their germination. The date on the herbarium sheet showed that the seeds were 237 years old. Since then, tough-coated seeds of the same plant have been found in Far Eastern peat bogs where carbon-14 dating has shown that

they have successfully survived centuries of dormancy. Of these seeds, the three oldest ones were recovered from a Neolithic canoe buried under 5.5m (18ft) of peat bog near Tokyo. Viability tests under vacuum conditions have shown that even many years' storage causes no change of viability in at least 50 different plant species, though even so certain species do tend to deteriorate. The final results of these tests, of course, will not be available for many generations to come.

Germination

For many plants, springtime brings with it the correct combination of triggers needed for germination. Some very short-lived plants, called ephemerals, can only exist in their very dry habitats (such as the tops of walls and sand-dunes) by virtue of their very early spring germination, enabling seeds to be set before the summer drought sets in. With the right conditions a seed will begin to germinate. Water uptake continues for a week or so, but its effects are no longer generalized. Enzymes are wakened into action, digesting starch to sugar and releasing growth factors, respiration is stepped up with a massive energy release, and all this is aimed at the tiny plumule and radicle, the young stem and root. The cells of these two organs now undergo a drastic change in form; they elongate by as much as 10 to 100 times their previous length, a feat made possible by the supply to each root and stem cell of vast amounts of sugar sent up from the cotyledons. Now the cells are able to take in great volumes of water, relative to their size, and this stretches

them 'beyond the point of no return'. The walls of these cells are so constructed that they can only get longer.

Suddenly the seed can retain its mass of frenziedly active cells no longer. The testa bursts, and out pokes the root which stretches down and anchors the seed-sprout firmly in the soil. Next the stem emerges, the manner varying from species to species. Sometimes, as in the sunflower, the stem carries the cotyledons above the ground. Here the seedling leaves turn green and start to help with food production. If the cotyledons are thick with food, as in the pea, they often stay underground. Unlike the root and stem, they cannot grow very much, their cell walls being thick and inflexible, and once they have enabled the plant to gain a foothold in its new environment, the cotyledons become redundant.

The period of germination ends with the development of the first foliage leaves; the little plant is now completely self-supporting and has reached the seedling stage.

Seed dispersal

An aspect of the life of a seed so far overlooked, is that of its dispersal—a matter of vital importance to the plant-to-be since it offers escape from the rather unsatisfactory fate of germinating *en masse* immediately beneath the parent plant. Few seedlings can survive such crowding by members of their own species, all struggling against each other for identical requirements in food, water and light. Since plants are sedentary, the reproductive phase offers the only opportunity in the life cycle of the plant to widen its range, to seek and exploit the potentialities of other habitats, with all the attendant possibilities for further evolution that these can bring.

Flowering plants usually produce large numbers of seeds as an insurance against failure of many to reach agreeable destinations. In this way, some at least will find suitable conditions provided that the area over which they are scattered is wide. Since, like pollen grains, seeds are not self-motile, plants have adopted a variety of means in order to effect this wide dispersal of their seeds, means both mechanistic and involving outside agencies—chiefly wind, animals and water. In many cases the dispersal of the seeds is closely connected with that of the fruit.

Wind dispersal

Plants with very small seeds may have these rather than their fruits blown away by the wind. Small size combined with lightness, often achieved by having a loose dry testa, can make such seeds as easily air-borne as spores. Many orchids rely on this method for the dispersal of their seeds which are smaller than those of any other plant and take to the air with great readiness. Other seeds develop specific structures to aid their wind dispersal. The milk-weed has seeds with tufts of silky outgrowths of the otherwise hard testa and these catch in passing breezes, carrying the seed on its way. In the rosebay willowherb (*Chamaenerion angustifolium*), the fruit pods split open whilst still attached to the parent plant, releasing seeds with long silky hairs which are blown out and scattered to considerable distances on account of these hairs. Sometimes winged seeds occur too; wings enable the seed to coast along in the breeze and are to be found for example in honesty (*Lunaria biennis*) and the field spurrey (*Spergula arvensis*). The conifers and the yellow rattle have winged seeds too, while willows and poplars have woolly seeds.

Of all the wind-dispersed seeds, the small-and-light ones tend to be by far the most widely dispersed, while winged seeds cover the least distances, though winged seeds of the Scots pine have been recorded as travelling 792m (880yd). Wind-dispersed species are often the first to colonize open patches of ground because their seeds are generally so readily transported and randomly deposited. A good example of this is the rosebay willowherb, also popularly known as fireweed on account of its rapid appearance on fire-devastated areas.

Animal dispersal
Generally speaking, animals are more important in the dispersal of fruits rather than seeds alone. This is because seeds lack the immediate food appeal which fruits hold for many animals. Sometimes birds feed on seeds. Of course, should a seed become broken up in the bird's beak or gizzard, it will no longer be capable of germination, but it is an interesting fact that the intestines of a seed-eating bird such as a finch will usually be found to contain many intact seeds which will be able to germinate upon their exit from the bird's gut.

Quite a few plants, especially woodland species, have their seeds dispersed by ants. Such seeds typically possess little structures impregnated with an oily food-substance which ants find attractive. This structure, often brightly-coloured to increase its appeal to the ants, is called an elaiosome, and is found for example in the gorse. Ants carry these seeds off to their nests, dropping some on their way, and thus effect dispersal.

As far as dispersal by carriage on animals' bodies goes, fruits generally show more widespread structural adaptations to this method than seeds. Beggar's-tick, cocklebur and stick-seed all possess hooks and spines for clinging to fur and feathers.

Water dispersal
To be dispersed by water, seeds must be so adapted that they combine buoyancy with an ability to retain their vitality upon submersion. Each seed of the white water lily *(Nymphaea alba)* possesses an air-trapping spongy aril in addition to a testa with many air pockets. The seeds are liberated underwater, and they float away in the water currents. After a while, the seeds become waterlogged and sink to the bottom of the stream where they germinate.

Mechanical devices
The seeds of many plants are dispersed by forcible ejection by the fruit. Touch-me-not and jewel-weed have seed pods which swell as they mature so that when the pods finally burst, the ripe seeds enclosed within are hurled out to some distance. The witch-hazel fruit splits open slowly, squeezing its slippery seed between its moist halves, and finally sending it skidding on its way. *Ecballium elaterium*, better known as the squirting cucumber, has a very turgid fruit which, upon dropping from its stalk, leaves a hole in its lower end through which contraction of the fruit 'skin' releases pent-up pressure forcing the seeds along with the excess fluid out through the hole to a distance of several metres.

The poppy, *Papaver rhoeas*, disperses its seeds by the 'censer mechanism'. Here mechanical dispersal is assisted by the wind. The ripe fruit develops a ring of holes around its upper side; through these holes, the seeds are jerked out like pepper from a pepper-pot as the wind whips the fruits about on their long tough stalks. Many other plants, amongst them the campion, primrose and monkshood, also use the censer mechanism.

Gorse combines another method with mechanical dispersal. As its fruits dry out, tensions are set up which eventually burst open the fruit in a twisting action accompanied by a loud pop, causing the seeds to be shot out. These seeds, lying on the ground, are then picked up and dragged off by ants, as mentioned earlier.

Generally speaking, mechanical dispersal does not usually succeed in spreading the seeds to any great distance, and is therefore not really a very efficient dispersal method. Perhaps this explains why so many plants do in fact combine other agencies with this method, as in gorse and poppy.

Accidental dispersal
Quite often seeds are dispersed accidentally, by-passing the special method for which they have become adapted. Any seed which will float can be carried off by water and, provided it does not remain wetted for too long, it may survive the journey. Seeds of water plants specially adapted to floating may become stranded on mud and then stuck to the feet of ducks and other water birds and transported in this way from one patch of water to another. Any kind of seed can be blown immense distances by such things as freak hurricanes. Grasses have light seeds which are normally wind-dispersed but one species, called *Cynodon*, is regularly dispersed in rather a strange fashion in the Belgian Congo. Here termites store large quantities of the grass in their nests and, when these are later abandoned, the grass seeds have germinated and established themselves there.

Perhaps we can go even further and say that some species are actually adapted to be non-specialized, and have come to rely on 'accidents' for their further existence. For example, the rayless mayweed *(Matricaria matricarioides)* appears to have no obvious adaptations to dispersal, and yet is widespread and extremely common on roadside verges. Other members of the family, the Compositae, have fruits specially adapted to wind dispersal. It appears that this plant relies in fact on carriage by motor car tyres—a kind of plant hitch-hiker. In days before the internal combustion engine it presumably spread by way of cart-wheels and boots.

Man ranks high as a chance dispersal agent. His influences in this respect can be seen everywhere, both purposeful and accidental; seeds of many kinds have been transported into practically every corner of the earth, resulting in the establishment of species in places which could never otherwise have been reached. The red-top grass was introduced into New Zealand quite unwittingly by man: emigrants travelled to New Zealand from Nova Scotia with mattresses filled with hay which included red-top. Upon settling in the new country, the mattresses were abandoned and the seeds of this American grass germinated and became established in this their newly adopted country. Protected motorway and railway verges, and canal systems, provide plants with a continuity of habitat over vast areas, acting as an uninterrupted network of corridors along which plants can spread. By such indirect means, man unknowingly aids the dispersal of many plants.

Right: Many seeds owe their spread to being eaten in fruits and passed out unharmed in excreta. These tomato plants are growing in sewage sludge after being eaten by humans.

FRUITS
for protection and dispersal

The ovary undergoes special changes as as a result of fertilization, gradually forming that familiar part of the flowering plant—the fruit. Essentially the fruit is the container for the seeds, and it develops side-by-side with them, affording protection from both mechanical damage and the effects of the elements. By the time the seeds are mature and ready to be dispersed, the fruit is ready to play its part too: the ripe fruit is specially adapted to secure dispersal by one method or another so that its seeds may be spread over as wide an area as possible. Sometimes, the pulp of the fruit contains chemical substances which control the dormancy of the seeds. This occurs in the tomato, pear, apple and lemon, and the seed, to escape its dormancy, must be physically separated from the fruit. The fruit is thus a very important structure, and the success of the flowering plants, so obvious today, owes much to its existence.

'True' fruits and 'false' fruits

The word 'fruit' really only applies, in the strict botanical sense, to the structure resulting from the female part of the flower, the gynoecium, alone. A true fruit would thus consist of a much-enlarged ovary with perhaps the remnants of the style and stigma, shrivelled but still attached. However often other floral parts, besides the gynoecium, have gone into the make-up of a fruit. For example, in the apple the fleshy edible part is in fact developed from the receptacle of the flower, and the true fruit lies further inside, forming the core with its pips being the seeds. This fruit is called a 'pome' and is typical also of the hawthorn, quince and pear. A much-swollen receptacle is also responsible for the fleshy part of the strawberry, in this case the pips being the true fruit. The fig is a strange and complicated structure. A single fruit is derived from a mass of little flowers which become covered by a succulent container formed from the inflorescence stalk. The pineapple again includes a whole inflorescence on a fleshy stalk in its massed and succulent fruit (all the floral parts being included). Such fruits are called false fruits, or pseudocarps.

An interesting point concerning the greengrocer's distinction between fruit and vegetables is that many of his so-called 'vegetables' are in fact true fruits, such as the tomato, the marrow, French and runner beans, and the cucumber.

There are two principal lines along which a fruit may develop. One leads to the formation of a hard dry tough exterior, the other to a soft fleshy succulent form. The exact form depends largely on the original structure of the ovary, which can vary considerably.

Berries and drupes

Fleshy fruits are of two major sorts: berries and drupes. The ovary wall or pericarp of a berry is fleshy all through, as in the tomato. It has a firm skin called the epicarp surrounding the central succulent mass, which is the mesocarp. The seeds lie in the centre of the berry, being the 'pips' in the tomato. The banana, too, is a berry. It usually will be found to have no seeds because it is commercially developed that way, but in the centre of the fruit are some small brownish objects—the remains of the ovules. Citrus fruits, such as oranges, lemons and grapefruit, are berries in which the mesocarp as well as the epicarp has become firm and leathery, and the carpels themselves, fused in a compound ovary, form the fleshy edible part. If the skin (epicarp) and the bitter white part under the skin (the mesocarp) are peeled away the juice-

Capsicum annuum
Sweet Pepper

Right: The orange is one of several citrus species with fleshy pulp, probably originating in China, which man has developed over thousands of years into hundreds of varieties.

packed carpels can be separated out with a knife.

A drupe, on the other hand, has its pericarp in three layers, not two. The epicarp is firm, the mesocarp fleshy, as in the berry, but the third layer, the endocarp, is very hard and stony. This endocarp surrounds the typically single seed, as in the plum, a typical drupe. The plum stone is not the seed: it is the stony endocarp, and when it is broken open, the real seed is inside. The walnut *(Juglans regia)* and the coconut *(Cocos nucifera)* are both drupes, but they are sold in shops with the epicarp and mesocarp stripped off, leaving only the hard endocarp. The nut kernel is the seed. Quite often a lot of little drupes are gathered together on a receptacle to form a fruit such as the raspberry or the blackberry.

Berries and drupes constitute an important part of commercial fruit production.

Dry fruits

In many fruits the pericarp does not become fleshy; instead it becomes toughened and dry, and sometimes woody as well. When they are mature these dry fruits may split open to disperse their seeds (dehiscent) or they may not (indehiscent). When a single carpel enclosing a single seed becomes an indehiscent fruit, it is termed an achene. Examples of achenes are found amongst the buttercups, the wood avens *(Geum urbanum)* and clematis.

A nut is usually an achene too. A hazelnut, for example, has a hard woody shell, the pericarp of the fruit. Now, if the nut is cracked open, out falls the single seed, the part which is eaten. Many other nuts are achenes too, including the fruits of the sweet chestnut *(Castanea)*, beech *(Fagus)*, and oak *(Quercus)*. Exceptions are the walnut and coconut which are technically drupes.

Most achenes have a separate seedcoat and fruit wall, but sometimes, as in the grains of cereals and other grasses, the two become fused together, forming what is known as a caryopsis. Other achenes have their pericarps produced into a wing as in ash and elm, whilst the sun-flower, being typical of its family, the Compositae, has a sort of false-fruit-achene as its pericarp is partially derived from the receptacle. A fruit formed from collections of little achenes is quite a common type; it can be found in the buttercups and also in the strawberry where the achenes develop as the little 'pips' on the expanded receptacle.

The forms of dry dehiscent fruits are varied. The capsule, a common type, develops from carpels which are fused together (a syncarpous gynoecium) and opens by splits, pores, or teeth, as in many of the wallflower family, the Cruciferae.

Distinctions between a fruit and a seed

Isolated from the plant, a fruit is sometimes difficult to distinguish from a seed. Remember that a fruit is a fully-ripened gynoecium, which should therefore hold seeds inside it, and look for these. On the outside, a fruit will usually show the flower stalk by which it was attached to the plant, whereas the seed will show a scar called the hilum, the position of its detachment from its ovary stalk. The withered remains of one or more styles are visible on the fruit, but these are never to be found on a seed. Sometimes a seed shows a little hole or pore, called the micropyle, but no such hole is borne by a fruit. Upon opening, the differences revealed are more striking. A many-seeded fruit has its pericarp enclosing the seeds, which are separate from each other and the pericarp. Difficulties may arise where the fruit is one-seeded, as in an achene, particularly if it is also very small. The outer covering(s) has to be peeled off to display

Right: The anciently cultivated quince **(Cydonia vulgaris)** *makes a large fleshy fruit which contains and protects the seeds.*

Below: Holly berries are red to attract birds, which enjoy the fleshy pulp. Many of the small seeds are unharmed when the fruit is eaten, and are voided by the birds at a distance from the parent tree.

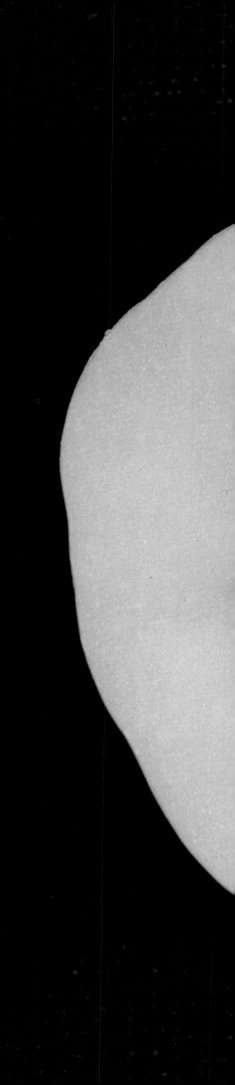

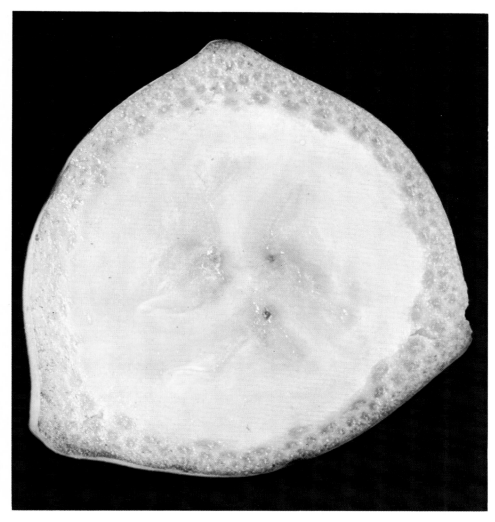

The banana is a fruit now grown all over the tropics; carried in large clusters (left), it is sweet, fleshy and nutritious, and millions of people subsist upon it. Although wild bananas have normal seeds, the cultivated form lost them (above) at some stage thousands of years ago, and this seedless form was seized upon by man who has developed numerous cultivated varieties, which are increased from suckers.

the embryo. One layer usually indicates a seed, though the caryopsis, with its fused pericarp and seedcoat, must first be ruled out. If two layers have to be removed, it is safe to assume that it is a fruit, the two layers being the fruit pericarp and the seed testa. Sometimes a third layer is present. This is probably due to a cupule which surrounds the fruit until it is mature, and in such a case would be responsible for the outermost layer.

The fruit as an agent of seed dispersal

Although the fruit serves first and foremost to house and protect the developing seeds, it also plays a secondary, but also very important, role of securing the dispersal of the seeds. In most plants it is the fruit which is the main agent of dispersal.

Wind dispersal

Wind dispersal is very common, both among fruits and seeds, and there are numerous modifications to be found associated with this method. Both fruits and seeds can develop a very similar mechanism for aiding wind dispersal. Wide dispersal can be achieved, for instance, by having very lightweight fruit, like seeds, which can take to the air with great ease. Such fruits are often called 'dust fruits', and a good example of this is the exotic wormwood, *Balanophora*. Other fruits become winged, plumed or woolly, again like seeds. The fruit of the elm *(Ulmus)* is an achene, and its pericarp grows out to form a membranous wing. This can act as a sail or propeller to catch the wind. In the ash *(Fraxinus excelsior)* the fruit is a pair of achenes, each of which has a long twisted wing which sends the fruit spinning on its way. The sycamore and the maple are also similarly dispersed. Some fruits have wing structures which develop from parts other than the carpel wall; in the hornbeam and hop fruits, wings develop from bracts, while in species of *Salvia* the calyx may form a sail. The fruit of the dock is covered by the calyx which develops three wings.

Usually the stigma and style of a flower wither away soon after fertilization but sometimes they can persist and come to aid in dispersal. In traveller's joy *(Clematis vitalba)* the style remains and becomes feathery to catch in passing breezes. This plant is called 'old man's beard' by some because of the grey fluffy appearance of its fruits. Many members of the Compositae, the daisy family, develop a ring of hairs, called the pappus, in place of the calyx in each fertilized floret. The dandelion develops a long stalk on top of the ovary to carry the hairy pappus at its end, whilst the remainder of the fruit develops into an achene. A ripe dandelion 'clock' thus consists of a collection of little

achenes, each with their pappus and stalk, positioned on the flower base, a structure which starts out as being flat, but curves in drying so that the fruits assume a spherical arrangement. This positioning is ideal for dispersal by wind, the slightest puff of which lifts and carries the fruits away like tiny parachutes, often to great distances.

Woolly fruit, for example the fruit of the pasque flower, function in a similar wind-catching manner to those with wings and plumes.

In steppe and other areas which tend towards desert conditions, a type of wind-dispersed plant known as a tumble-weed is common. Here, where vegetation tends to be low-growing and wind velocities high, the whole plant or a large chunk of it can be blown along by the wind as a regular method of dispersal. Elsewhere, some kinds of grasses and seaside plants such as the sea bladder-campion *(Silene maritima)*, constantly subjected to strong winds in an open habitat, are dispersed by a slightly less extreme version of this tumble-weed method in which the seed-bearing flower head in its entirety is blown off by the wind.

Animal dispersal

Dispersal of fruits by animals is very common and occurs either by the animal eating the fruit and later voiding the seeds, or by the fruit being carried off on its feet or hide. In fleshy fruits, the succulent portion is usually specially modified for catching the animal's eye. Until ripe, the fruit stays green, merging inconspicuously with the foliage. As it ripens, the fruit develops a bright conspicuous colouration, frequently in shades of red to contrast with the green leaves. At the same time the fruit develops its characteristic appealing odour and attractive texture. These fruits attract a variety of animals. In temperate regions, birds are the chief consumers, but in the tropics, monkeys, wild pigs and fruit bats are amongst the assortment of creatures which visit plants for their fruits. In feeding, the animals either discard the seeds, well-protected in their tough seedcoats as in berries and achenes or inside stony endocarps of drupes, or swallow them. In the latter case the seeds pass through the gut system and out

Pyrus communis
Pear

with the excreta, generally unharmed and sometimes in a better condition for germination. Quite often a bird will gorge itself on its favourite fruit, fly away and then eject the pulped mass, apparently suffering from the effects of over-eating!

Fruits can be carried externally on an animal's body, and common adaptations to this mode of dispersal are hooks and spines, bristles and mucilage, any of which may cause the fruit to stick to fur or feathers of a passing animal. Almost any part of the fruit wall can become modified in this way; in the wood avens and the buttercups, it is the style which becomes hooked, in the carrot, woodruff and goose-grass, it is the ovary wall, whilst in the forget-me-nots (*Myosotis* species) and bur-marigolds (*Bidens* species) the fruiting calyx develops hooks.

In plants which rely on animal dispersal, we find the dispersal structures developed are generally fewer and less complicated than those developed as adaptations to that other major dispersal agent—the wind. This is because animals tend to frequent areas which are fertile, so that fewer seeds end up on barren ground. The plant can thus 'afford' to minimize its expenditure on reproductive devices. According to the type of animal involved, dispersal of fruit by animal means can be quite far-ranging. This is particularly true with birds, and also with the fruit bats which have been known to travel 48km (30 miles) on nocturnal fruit-foraging expeditions for mangoes, guavas and bananas, returning to their roosts the same night.

Of course, animals can also disperse fruit in other rather less-conventional means. Hedge-hogs can accidentally spear fleshy fruits onto their spines and carry them considerable distances before dense undergrowth brushes them off again. Nut-eating animals like squirrels can make a cache and never find it again. Birds and other animals can carry fruits

Above: The enormous jak fruit (Artocarpus integrifolia), weighing up to 80kg (176lb) is a very primitive fruit that drops straight off the tree. Its starchy pulp is eaten by animals which help spread the seeds.

Right: Hundreds of small separate fruits coalesce to form the fleshy pineapple (Ananas comosus). Man has done much selection to improve fruit size and also reduce spininess of the foliage.

stuck to their feet. The Malayan plant *Rafflesia arnoldi,* a parasite on the roots of vines, has its fruits dispersed by elephants; the fruits are large and squashy so that they stick to the elephants' feet and the seeds are carried and pressed down against the roots of the host plants. It seems that only an elephant has a

Mutiny on the Bounty

In 1787 the 215-ton vessel **Bounty**, commanded by William Bligh, left England for the specific purpose of transferring live breadfruit trees from their native Tahiti to the West Indies. A gardener from Kew, David Nelson, was responsible for collecting and growing the plants in special quarters. However, after eventually leaving Tahiti, the ship's crew mutinied: Bligh, Nelson and 17 others were set adrift in a small boat, and the 1,000 breadfruit plants were thrown overboard. Though Bligh has been portrayed as a hard captain, the crew most probably rebelled against the occupation of limited cabin room by the outlandish plants and the unnatural work of tending them, including the use of precious drinking water. After a journey of 3,518 miles, Bligh brought the castaways to Timor, where unfortunately Nelson died of a fever. In 1791–3 the breadfruit expedition was repeated, again under Captain Bligh with two Kew gardeners, and a shipload of breadfruit and other useful Tahitian trees was safely brought to Jamaica.

tread which is heavy enough to press the seeds into the soft jungle soil where they can germinate up against the vine's roots.

Other methods of dispersal

Many plants have self-dispersal mechanisms, but the fruit itself does not become dispersed with the seeds.

Dispersal of fruits by water is rather uncommon. Many fruits, as with seeds, will float but they do not remain viable for long under such conditions. Adaptations for water transport in the fruits of those plants which do rely on this method take the form of loose or spongy coverings which can retain pockets of air to give them buoyancy. The yellow water lily (Nuphar lutea) has a large berry fruit which, when ripe, splits into its component carpels, each of which is lined with mucilage containing air bubbles so that they can float away. Later the carpels decay, releasing the seeds which then sink to the bottom of the water where they germinate.

Perhaps the best example of a water-dispersed fruit, however, is the coco de mer or double coconut of the Seychelles (Lodoicea maldivica), which is one of the largest fruits known. This fruit, along with that of the coconut (Cocos nucifera), has a tough woody covering to protect the seed and a lot of air-trapping fibrous matting to give the fruit buoyancy during its passage in the ocean currents. These plants are often amongst the first to colonize oceanic islands or coral reefs because of their readily water-borne fruits.

Those plants which grow totally immersed in water rarely produce buoyant parts; many produce achenes which sink quickly on being shed.

Fruits and man

Of course, fruits can be dispersed by chance, by-passing the 'set' methods available by animals, wind or water. Not the least of these chance agencies is man, for whom fruit has held a strong attraction the world over since Adam and Eve. Too strong an attraction for some, perhaps, since the deaths of Pope Paul II in 1417, the German King Frederick the Peaceful in 1493 and his son Maximilian in 1519, were all supposedly caused by a surfeit of melons. In AD 200, the Graeco-Roman 'prince of physicians', Galen, expressed the opinion that his father had lived to reach the age of 100 because he never ate fruit. Deprived years, we might well consider today, when fruit plays such an important part in our diet, both from gastronomic and nutritional points of view.

All over the world, fruits of many kinds are being cultivated on a large scale. Many of the wild ancestors of our cultivated fruits bear fruits which, by today's standards, seem small and inconspicuous in comparison. The wild banana (Musa rosacea), for instance, bears banana fruits which have little pulp and are barely edible. Over the years careful selective breeding has produced plants bearing bigger and better crops than ever before, and varieties to suit almost every kind of demand, which are often seedless into the bargain. This is particularly true of that most beloved of all fruits—the apple. According to Genesis, the apple (Malus pumila) was created three days before man, but, unlike man, it has never fallen from grace; today it is the world's most cultivated tree fruit. Varieties of apple abound; some of them have become deservedly famous, such as the Cox's Orange Pippin and the Golden Delicious.

Besides being a pleasure to eat, fruit also provides an important source of vitamins and minerals in the human diet. British sailors on long voyages used to suffer from scurvy, a severe skin disease caused by a vitamin deficiency until it was discovered that upon supplementing their restricted diets with fresh citrus fruits, the symptoms disappeared. Hence the British sailors came to be known by the Americans as 'limeys' on account of their habit of hoarding fresh limes on board during long voyages. Large amounts of vitamin C provided by citrus fruits are thought today to give resistance to the common cold.

Vegetative propagation of plants

In some plants, reproduction by means of fruits and seeds produced by the sexual process does not occur, and the only way in which they can manage to reproduce themselves is by the so-called vegetative method. Whole new plants grow from plantlets or bits and pieces derived from an already established plant. Since no genetic variability is involved, these new plants are just like the parent plant in all their characteristics. Sometimes plants are propagated vegetatively on a commercial basis because of this feature, as it enables a desirable character to be preserved whereas with seed planting, it could be lost in genetic juggling. In nature, this vegetative propagation results from several methods. In the grasses and cereals it results from the rooting of stems from tillers whilst in the strawberry, runners are produced. Other plants use all sorts of bits and pieces. The Canadian waterweed can grow from any part severed from the parent plant and consequently can rapidly establish itself over large areas of water. Irish potatoes are grown from pieces with eyes, which are tiny dormant plant shoots, whilst in horse-radish buds develop from pieces of root. Buds also develop on leaves in some plants; a whole African violet plant can be grown from just a leaf with its base or stalk placed in water or earth.

Right: All kinds of animal help to disperse the seeds of plants. The seeds of the tasty blackberries this dormouse is eating are tough enough to pass through its gut unharmed and germinate after excretion.

Below: The bitter gourd of the desert (Citrullus colocynthis) separates from the dry stems when ripe and the wind bowls the 10cm (4in) fruits across the sand. They eventually burst open in the sun.

PLANT COMMUNITIES AND ADAPTATIONS
filling the niches

In nature, plants are seldom found growing with only their own species; it is usually man's interference by cultivation which excludes all other species. Most plants will be found growing in the wild as part of a heterogenous community. These communities may be extremely complex or very simple, but in each case they are entirely suited to the immediate conditions. The structure of a plant community is determined by several environmental factors: climate, soil conditions, humidity, seed availability and also on such conditions as seed viability and methods of plant reproduction. Not only do these factors act individually, but they also interact, thus ensuring that each community consists only of plants which are adapted to survive the prevailing conditions.

As may be expected of nature, few communities are clear-cut; there is in fact considerable overlapping and gradation. Recognizable plant communities do not occupy sharply defined areas, but merge through zones of competition where typical species of one community intermingle with those of another. These zones of transition are known as ecotones. There is even a school of thought which denies the existence of plant communities, preferring the 'vegetation continuum' theory, which measures vegetation as a whole, varying in both time and space. The main disadvantage of this latter theory is that it requires an infinite number of measurements to account for all of the variations before the vegetation can be described. Although this may be the ultimate ideal, it is impossible to carry out. Therefore the plant community is recognized as the acceptable method of vegetation classification.

Classification of plant communities

Plant communities are commonly classified by one of two methods, that is either by the dominant plant form, or by the species present. The first of these two methods is favoured by plant geographers, as it is common for one plant form or growth habit to dominate the vegetation of an area. Perhaps the best and most easily recognizable example of a growth form dominating a geographical region is the forest. Here there is sufficient energy input to the community in the form of moisture, heat, light and nutrients to support the arboreal growth form with its great demands on ecological resources. In direct comparison, desert areas with low-energy input will only support small slow-growing plants.

The second form of classification is possibly the more obvious of the two: communities with the same floristic composition are the same. Unfortunately, this statement is never true, as no two identical plant populations will ever be found. When it is considered that so many interacting factors are responsible for the make-up of a plant community, it is understandable that the chances of identical conditions existing are negligible. This fact does not render this method useless, however, as it is of great importance to know what species are present; it merely means that floristic composition only enables classification into community type. Well-known community or vegetation types are salt marsh, freshwater marsh, chalk grassland, peat bog, desert, forest, tundra and many others.

Further classification difficulties arise when the ecology of an area is so varied that a patchy vegetation results. Marshland would come into this category, where the whole may be divided into wet dips with a completely different vegetation from the neighbouring dry hummocks. In such cases each unit is called a microstand, and together they make up a community complex. Without such limits being imposed, vegetation could be divided into such small units that the word community would become meaningless.

Factors affecting distribution of communities

The distribution of plant communities is determined by a great number of inter-related factors. Most obvious, perhaps, are those relating to climate, including temperature, precipitation, sunlight and average wind speed, but also of great importance are the physical and chemical composition of the soil and thus the soil parent material. An area with a well-drained soil and a heavy rainfall may support a rich forest, whereas the same rainfall on a less porous soil may result in a peat bog. Thus as the geology and climate vary across the earth's surface, so the vegetation differs.

The communities vary as the plants within them adapt to the climatic conditions, until ultimately a climax community is reached. Such a community is basically the highest form of plant life which may be supported by each environment. Complete stabilization may be

Within certain limits, plants will adapt to external climatic conditions, like this wind-pruned hawthorn, shaped by the prevailing coastal wind.

The Galapagos Islands are the only place in the world where the prickly pears (above) develop tree-like trunks complete with bark. It has been surmised that they have evolved in this tree form, with their succulent stem-pads well away from the ground, to protect the latter from the giant tortoises (left) which inhabit these remote volcanic islands. They also have especially sharp stiff spines.

said to exist only when the climax is dominated by trees. This is because trees are the most demanding of their environment of all plants, and it is considered that certain forest communities utilize all of the energy for plant growth available in that environment.

Obviously not all vegetation has stabilized to a climax; most reasonably stable communities are called subclimax. This temporary stabilization is caused by a certain arresting factor, either natural, or as a result of man's actions, which prevent further changes. A typical example of a subclimax community is heathland, where *Calluna* and *Erica* species are dominant on what is usually an acid sandy soil. The climax vegetation for such a soil in the temperate regions would be a pine or birch forest, but this is prevented from forming by fire—the arresting factor in this case. In the dry summer conditions on the heath, fires are common, the shrubby heather bushes can withstand burning, and even if almost all growth is destroyed they will soon shoot again at surface level. The young tree seedlings, however, which

will have just germinated between the heather bushes, will be completely destroyed, thus preventing the establishment of a forest.

Under perfect conditions, vegetation develops from an original pioneer community through what may be a succession of subclimaxes to a climax community. Such successions are called seres and various types exist. Successions on dry land are called xeroseres. These may begin with the colonization of rocks by microscopic algae and lichens which gradually break down the rock to form a 'soil', allowing colonization by slightly larger lichens. This process continues until there is sufficient soil to support higher forms of plant life. The xerosere may then continue through subclimaxes of perhaps grassland and hazel shrub to a climax of oak forest.

Hydroseres are similar successions, only they start under water. First, small water plants become established near the edge of a pond, silt and plant remains collect around these, leading to a build-up of soil and a consequent shallowing of the water. These changes enable reeds and bulrushes to colonize; with a further build-up of soil material around their stems. This is the beginning of the formation of peat, and in an alkaline water area this would probably then become fen, and shrubs and trees which can tolerate a waterlogged soil would eventually take over.

The vegetation of any one area varies both in time and space. The variation in space is easily seen by noting the obvious differences in the vegetation over the earth's surface. The variation in time is the seral succession as

described above. In both, the community figures greatly, and is obviously an important concept which must be considered carefully when studying vegetation.

Adaptation of plants to prevailing conditions

Communities are composed of plants which are adapted to the existing conditions. In fact, plants must be adapted to their environment if they are to survive, and it is this plasticity which is one of the most unusual and fascinating features of the members of the plant kingdom. Adaptations allow some plants to withstand drought, while others can survive waterlogging; some to thrive in extreme cold areas, and yet others to withstand great heat.

Several experiments have been undertaken to determine to what extent a species may adapt. For example, numerous transplants were made along a line across the Sierra Nevada mountains. The experiments were very carefully controlled, ensuring that only physical conditions changed, by transplanting each plant with its original soil. Identical plants were always used, often making sure that this was the case by taking cuttings from the same plant. They were then planted at different stages along the line; this meant that some were more or less at sea-level, while others were positioned at transplant stations upto 2,200m (7,200ft) above sea-level. The transplants were left to fend for themselves completely in their new and vastly different environments.

The results were staggering; some plants adapted to their new environment so well that after only a few years they would have been identified as a different species. Plants which had shown luxuriant growth in the lowlands became small and stunted at the higher levels. Their internodes were shorter, stems were more woody and leaves became smaller. In fact, the plants took on a typical 'alpine' growth habit. The plants on the lower slopes were found to be intermediate in form between the lowland and alpine types. Other scientists have repeated these experiments with different species, and this plasticity has been found to be quite common in the plant kingdom.

The ability to change the outward appearance is called phenotypic plasticity and it is very important for the survival of the plant. However, it is non-reversible; for example, in the above experiments, if the alpine plant were grown at sea-level it would not revert to its original form.

Plants of the same species exhibiting phenotypic plasticity appear different but genetically they are identical, and produce offspring of the same appearance. A plant becomes more perfectly suited to its environment when it is genetically adapted, as it will then produce offspring which are also adapted in the same ways. When variations and adaptations become fixed genetically they become characters of that species. This is Darwin's natural selection in action.

Most adaptations are now believed to arise by selective action of the environment causing genetic variation strictly by chance. Under such conditions, sexual reproduction becomes more important as it offers the chance to exchange genetic material.

Physiological adaptations

So far, only structural changes have been

mentioned, and at the beginning of ecological studies this was all that was considered, early ecologists merely speculating on the significance of these morphological adaptations. More recently, however, more significance has been applied to physiological adaptations. For example, it was at first assumed that the anatomical features of desert plants would reduce transpiration (water loss), but it has since been proved that some desert plants have a very high transpiration rate. These recent experimental findings have more or less halted the study of relationships between form and environment, and have led to a great upsurge in research into ecological physiology. A morphological change cannot be considered an adaptation unless there is a corresponding physiological change which makes the plant more suited to its environment. Needless to say, there is a lot more to learn about the relationships existing between plant form, function and the environment.

It is noticeable that the number of plant forms

Below: These tree euphorbias on the volcanic island of Tenerife display convergent evolution with cacti: both groups have evolved leafless water-holding stems to withstand dryness.

growing in any environment increases as the climate becomes warmer. This is thought to be because in a cold climate there is a very short growing season, and only species which can mature and reproduce in just a few weeks can survive. In a hot climate such as a desert, this selection for fast growth does not occur, and consequently there is a greater variety of morphological types to be found. These include succulents, mesophytic drought-evaders and highly drought-tolerant evergreens.

Quite different species in different environments will show the same adaptations. This is an example of convergent evolution, where two species have evolved the same adaptation independently. Evolution, after all, is merely plants becoming more highly adapted to their environment. It is believed that in many cases genetically fixed adaptations may outlive their usefulness, and this would explain such anomalies as succulent water plants. Once an adaptation has become a genetic character of a species, it cannot be discarded, although it may eventually disappear through mutation or natural selection if the plant reproduces sexually.

All plants are adapted to their natural environment, and man has made great use of this fact, although often unknowingly. The life cycles of

crop plants are carefully geared to fit the prevailing climate by planting at the right times, irrigating and protecting from frost. As modern technology has improved, even greater controls have become possible. Plant flowering is determined by the number of hours of daylight, and by artificial lighting it is possible to get flowers out of season. This ability to use plant adaptations allows the great variety of foods which we now enjoy all year round.

Forest communities

By examining certain communities in more detail, one may develop a better understanding of vegetation and how it adapts to its environment. A typical community which shows many forms is forest. Forests are made up of at least three strata of vegetation. These are basically the tree layer, below this the shrub layer and at ground level the herb layer. Although in most cases there are more non-arboreal plants than there are trees, it is the trees which determine the environment beneath them. On a world scale, the increasing severity of the environment is accompanied by a decreasing diversity in forest flora. In a tropical rainforest there are over 40 tree species per hectare (100 per acre), and there

The Saguaro desert community

Biggest of all cacti, the Saguaro, **Carnegeia gigantea**, sometimes reaches 15m (45ft) in height. Older specimens usually have several side branches, sometimes bent into extra-ordinary curves. Seedlings grow very slowly and large plants only put on about 4 inches a year. The biggest are over 150 years old. This huge structure has, like most cacti, an exterior of vertical ribs which carry the typical rows of spine clusters. These do not, however, deter many creatures from burrowing into the trunks. Chief among these is a species of woodpecker, which hollows a nest cavity from the pulpy stem. The damaged area soon heals over with a thick, hard layer of scar tissue, so that the woodpecker's nest is dry and snug.

When the woodpecker leaves the nest hole, it is very often taken over by the little elfin owl: this bird never digs out his own. Other small creatures like rats and mice may make burrows in the cactus stems, while yet others, like snakes, occupy holes made previously.

The large creamy flowers of the saguaro are followed by big pulpy fruits, which the Indians used to gather in great quantities for eating and drying for winter use.

may be as many as one hundred per hectare (250 per acre). Dominance by one or two species is rare unless man has disturbed the habitat. In direct comparison, within boreal forests, vast areas are found being completely dominated, often by just one species. These two types of forest are at the extremes of this community type, and are therefore probably the best examples to take to describe the diversity which may exist.

In equatorial regions the climate is continually hot and humid, and therefore provides the optimal conditions for plant growth. As a result not only are more species found, but also they tend to be larger. The tree canopy is usually composed of several substrata, each made up of trees of varying light requirements and height potential. Obviously those needing most light are the tallest and fastest growing. The shorter and slower growing species are positioned between their taller neighbours so as to obtain as much sunlight as possible. As growth conditions are so favourable, most tropical tree species are evergreen so that they can take advantage of the all-year growing season and not have a period of slow growth, as do the deciduous trees of more temperate regions. As the tree canopy is so dense (it can be up to 30m (98ft) deep), there is little light left to penetrate to the shrub or herb levels. This consequently limits the development of these layers to species which have become adapted to survive such poor conditions. It contains a large number of saprophytic and parasitic plants—these, after all, are the ideal adaptations to poor growing conditions. There are also large numbers of lianas and other epiphytic climbers which use the tall trees as a means of reaching the sunlight.

The temperate pine forests also have a very dense canopy, although here the lack of light penetration is also aided by the low angle of the sun in the sky. In natural pine woodland the trees are often found growing with birch, which has similar ecological requirements. In the single-species pine plantations there will be no shrub layer, and very few plants at ground level. Not only is there very little light here, but there is also a very poor soil as the pine needles have very thick cuticles, and these are slow to break down and add nutrients to the soil. The plants found here, again, will be saprophytes, parasites and epiphytes. Fungi are often the main species found, these do not need light as they are non-photosynthetic, and in any case they frequently have mycorrhizal relationships with coniferous trees in particular.

One of the best communities to show the relationships existing between the layers of a forest is the European temperate oak forest. Oak wood can exist under a variety of conditions, but the accompanying species will vary with soil depth and available nutrients, etc. In an area of high rainfall and with an acid soil, it would be common to find the ground beneath the trees completely covered by a dense growth of bilberry bushes (Vaccinium) accompanied by a thick growth of several moss species. Growing to a greater height, but shorter than the trees will be such species as mountain ash (Sorbus). Epiphytic lichens and ferns would grow on the oak trees, so that all available habitats would be exploited by suitably adapted plants.

Grassland communities

There are few types of vegetation in which grasses are not represented, and they have achieved an ecological dominance unrivalled by

The salt marsh system

Salt marsh vegetation is found on most muddy coasts in the temperate areas of the world. It is easily recognized as it exhibits a very definite community structure. It is made up of distinct zones of vegetation parallel with the shore line each with different plants from its neighbouring zone, although some overlap is inevitable. The best time to observe the salt marsh community is at low tide when all of the zones are visible.

On the mud flats nearest the sea will be green filamentous algae, very fine sea-weeds, which catch on to rocks and lumps of mud. Slightly higher up the marsh the same algae will be found draped around the plants along the shore, as they are carried up the beach at high tide. The first plant to root in the mud is usually the glass-wort (Salicornia), a small succulent plant highly adapted to its environment; it is tolerant of very high salinity, a constantly waterlogged soil and being covered by the sea at every high tide. The roots of Salicornia begin the stabilization of

the mud and allow other plants to become established. The first of these is probably the plant most commonly associated with salt marsh, the reed, **Spartina**. This plant is so successful in its habitat that despite the fact that it has only recently been introduced into Britain it is already found in most areas. It is very fast growing and its creeping growth habit means that it binds the shifting mud, very effectively stabilizing it to allow colonization by less highly adapted plants.

As soil build-up occurs in the upper regions of the marsh, there is a marked increase in the vegetation found and also in the number of halophytic (salt-tolerant) species. The reclaimed land is known as the emergent marsh, and is often marked by a small 'cliff' in the mud. The stable soil quickly establishes a mixed marsh community, namely salt marsh turf. The species present include sea plantain (**Plantago**), sea lavender (**Limonium**), thrift (**Armeria**), rushes (**Juncus**) and many others.

These plants are noticeably smaller and more compact in form than those of the lower marsh. They also usually have a smaller woody rhizome compared with the much longer creeping stolons of the **Spartina**.

The development of the marsh is marked by these zones. A young colony has no emergent marsh but nearly all the mud-flat plants. As the soil builds up at the top of the marsh, eventually non-halophytes will be able to colonize and the land will have been reclaimed.

Stratification of the rain forest

any other type of herbaceous plant. Grasses have an almost unique leaf shape, which coupled with their growth habit enables them to achieve the optimum photosynthesis for the light available. Their leaves are different from those of most other plants in that their growing point is at the base and not at the tip. This means that they can survive burning and still continue to grow. All grasses also have a large root system, far out of proportion to their shoot growth, which enables them to utilize to the fullest extent the available water and nutrients. They also produce large numbers of seeds on inflorescences which allow easy dispersal by the wind, and many also have forms of vegetative reproduction. All of these characters combined give the grasses a competitive advantage over most other herbaceous species with which they come into contact.

The grassland community structure is much simpler than that of the forest. It is dominated by the field or herbaceous layer. There are two main naturally occurring grassland communities, these are savanna and prairie. The tropical savanna consists of often quite tall grasses, although they seldom exceed 5m (16·5ft), growing with the occasional shrub. These communities are subject to heavy grazing and fires, which give the grasses the upper hand in the environment. The temperate prairie grasslands are controlled by the same factors, but the grasses found growing here are usually smaller. In both the prairies of America and the steppes of Russia, man has taken advantage of the grasses' natural dominance of the area, and has turned much of the land over to cereal production.

Adaptations to extreme conditions

Some vegetation has become adapted to survive the more severe conditions existing on certain parts of the earth's surface. Plants growing on the coast, for example, must survive high salt concentrations in their substratum, and often periodic battering by the waves. Salt-tolerant plants (halophytes) are also in the unique position of being surrounded by water which they are unable to use, because salt-water is of no use to plants. They therefore have to be adapted to conserve freshwater and to survive in water-logged soil which reduces the amount of oxygen available to their roots. Halophytes of the temperate regions are usually small herbaceous plants. By being small they offer little resistance to the wind and waves which batter their exposed habitat, and therefore are less likely to be damaged. They often have fleshy leaves and stems made up of large cells for storing water. It is also common that they have a reduced rate of transpiration as a further aid to fresh-water conservation. In the tropics, coastal vegetation is quite different; mangrove swamps are found here. The plants are large and woody with long aerial roots to support the bulk of the

Tropical regions with a high rainfall have a tree dominant vegetation. The trees are tall broad-leaved evergreens, in fact nearly all species are evergreen, as a deciduous plant would not be able to compete against species with an all year growing season. These forests are the most luxuriant form of vegetation, and have the fastest growth rate of any community. Some of the largest tree species on earth are found here.

The structure of the humid tropical forest is the same wherever it occurs; only the species differ. The basic requisite for this kind of vegetation is a hot wet climate. The mean monthly temperature does not drop below 25 C (77°F) and the relative humidity can be over 80 per cent. This humidity increases under the tree canopy with extra water being transpired by the vegetation. The annual rainfall is always above 1,500mm (59in) a year.

The tree-layer of the forest is usually made up of three different heights of tree. The basic canopy is about 30m (90ft) high, but there are also several taller species projecting from this layer, with some shorter ones growing below. This shortest group have narrow crowns to make the most of what little light penetrates this far. Below the trees is the shrub layer, the plants here are sparsely distributed as they only grow where light penetrates the canopy. The remaining layer that constitutes the ground cover is made up mainly of plants, such as ferns, that need little light.

Woody climbing plants called lianas are common, growing up the trunks and into the canopy to reach the sunlight. The trees also support a large number of epiphytes (plants attached to the trees and not rooted in the soil). Many different species have adopted this growth habit and they include ferns, orchids and bromeliads. These plants obtain their nutrients from decaying vegetable matter which they often trap in specially adapted cup-shaped leaves or matted spongy roots. It is common to find the lower parts of trees thickly covered with epiphytes.

The hot humid climate promotes the growth of this lush vegetation. It also ensures that any dead plant matter decays rapidly to add nutrients to the soil and thus enable it to support this large amount of growth.

The many kinds of trees and shrubs that make up a tropical rain forest grow in clearly defined layers, as this diagram shows.

Occasional giant trees thrust huge crowns of dense foliage into the airy light above the forest.

The broad crowns of closely-packed trees form an almost unbroken 'sea' of foliage 30m (90ft) above the forest floor.

Narrow-crowned trees compete for growing space in the light that filters between the trunks of taller trees.

In the patchy gloom beneath the trees sparse shrubs form a lower, separate layer.

The damp soil of the forest floor supports a rich growth of ferns and other shade-loving plants.

plant above the surface of the water. Having aerial roots also solves the problem of obtaining oxygen when growing in an oxygen-deficient waterlogged soil, as these plants can just as easily absorb this vital element from the air as from the soil.

Plants growing in an alpine environment have to survive some of the most severe conditions. They are often on exposed cliffs, can be subjected to extremes of temperature and usually have to grow on a poor shallow soil. They have overcome most of these problems by adopting the rosette or cushion growth habit and as a result they are little affected by high winds. They may often have shallow rooting systems which spread widely under the surface so as to obtain as much nutrient as possible from what soil there is. Most alpines have a short flowering period so that they may produce flowers even during the very short 'summers' in these areas. They are also perennials, so the species may survive when conditions do not allow annual reproduction.

Desert plants also tend to be perennials, as annual reproduction is not always possible, because of insufficient water. This same factor has led to the evolution of several forms of xerophytic plants. To conserve water, plants have developed numerous adaptations. Many have become succulents, composed of large cells with the ability to store water, for example, the cacti. Some have stems which curve inwards to trap water, and others have hairy surfaces for the same purpose. Other adaptations serve to reduce the amount of water lost from the plants, including the development of smaller and fewer stomata (the microscopic pores on the leaf through which water is lost and gases are exchanged), and slower rates of transpiration. As water loss occurs through the leaves it is usual for these to be small and in many cases they are reduced to mere spines, as in cacti.

Hydrophytes are adapted to a completely different environment, as these are plants which live in water. They must in some cases even be capable of obtaining their oxygen and carbon dioxide from the water in which they grow. In still or slow-flowing water, the problems are slightly less as here there will be some build-up of silt in which the plants can root and from which they can obtain nutrients. For example, the water lily (Nymphaea) is rooted in the silt and has its leaves floating on the surface of the water. In a fast-flowing stream there will be no soil, only rocks, and any plants found growing here are often algae which attach to these rocks and are filamentous so as to move with the water and not become damaged or dislodged. Other plants such as Lemna, the duckweeds, merely float on the surface with small roots hanging down into the water.

Man's influence on plant communities

Few communities seen today are completely

unaffected by man, who has both influenced established communities and created new ones. The most interesting for the ecologist are those which he has created, as here they have a chance to see the primary colonization of land and the subsequent development of vegetation. Man has created large barren areas with his slag heaps, and only by applying his knowledge of natural communities has he been able to vegetate them. The railways have added much to the richness of plant communities. Trains are the perfect agents of seed dispersal, and when the line is disused the sheltered areas remaining have a very rich flora. Where the face of the earth has been scarred by gravel workings, the usual result is areas of barren lakes. These can eventually be colonized, supporting a very varied plant and animal population. Unfortunately, there are also many results of industry which will remain as scars on the landscape for some time. They include the waste from chemical industries and slag heaps containing heavy metals. Here the adaptations to enable plant growth are so great that it will take many years for even a few plants to become established.

Plant communities exhibit a surprising complexity. Trees do not just grow on a certain hillside because there have always been trees there, but because at that time and with the

Above: This glade in a conifer forest is the result of animal influence; beavers have dammed a stream which has silted up over the years, providing marshy ground favoured by colonizing willows.

Right: To make their dams beavers cut down trees by gnawing and then gnaw the trunks into manageable lengths.

prevailing conditions, trees are the best-suited vegetation. This may not always be the case as the environment is constantly changing. The study of plant communities has changed dramatically in recent years. Originally plant ecologists merely observed and described what they saw. Now it is realized that the dynamics of the community are more important. Before a description can be drawn up, one must ask how the plants involved co-exist. The modern trends in this field now have a strong mathematical bias, as by using figures one can standardize the method of description more easily.

Plants are the basis of each food chain, thus plant life is important to us all. Perhaps if more were known about how plant communities work we may be able to solve some of our food shortage problems. The plant community is the basis of all plant life, its understanding is the culmination of all botanical studies.

Colonization of volcanic lava flow

One of the most destructive natural disasters is the volcanic eruption, the red hot lava destroys all it comes into contact with, no living thing in its path will survive. Yet volcanic islands can have some of the richest and most diverse floras in the world and in Italy some of the richest farm land is on the slopes of Mount Vesuvius.

Hawaii is hundreds of miles from the nearest land, and yet this volcanic island has a very rich and varied flora. By studying the plant colonization of volcanic lava here it was seen that no decomposition of the rock to form soil is necessary before plant growth becomes possible. As it cools the lava forms a hard porous rock which is rich in various minerals, especially phosphorus, and is capable of holding sufficient water to allow plant growth. On Hawaii, Kona, one of the largest forests, grows on a minimum of soil on relatively undecomposed lava flows.

Hawaii is so far from the nearest land that it is highly unlikely that even those seeds adapted for wind dispersal were carried this far by the wind. Most seeds were probably brought here originally by birds, but all species will have arrived through necessity by long distance dispersal. In fact as many as 50 species are known to have come from North America.

Some of the first species to have arrived will have been those with small fleshy fruits, such as **Vaccinium**, which are popular with birds. The birds eat the fruit before migration and deposit the seed in their excretion on the lava. Other early colonizers would have been ferns, their small light spores easily catching in the mud on the birds' feathers and feet.

Seeds may arrive by various methods. The largest single-seed fruits, such as the coconut, are capable of floating in sea water for some considerable time before they may be washed up on an island's shore. Other seeds may float on rafts or drift wood.

Such a fertile substrate as a lava flow will not long be without vegetation wherever it is situated. Even in the very arid Galapagos Islands a few plants grow on pure lava, including the endemic cactus, **Brachycereus nesioticus**.

PLANTS AND MAN
our green slaves

From the beginning of history man has had an economic interest in the plants which form part of his environment. This interest has been essential for his survival and subsequent development from the earliest civilizations to our present urbanized societies. About 10,000 years ago agriculture was born; man learned to recognize specific plants and animals as being particularly valuable and started, albeit in a rather primitive way, to encourage their growth by planting, feeding and protecting them. In principle it was a short step from this to the more rigorously controlled plant and animal husbandry that is familiar to us today. As man's knowledge of plants and plant products has increased so has their use and exploitation and there are hardly any aspects of any present-day society which have not been influenced in some way by the use (or occasionally misuse) of a plant product.

Plants as a source of food
Cereals and vegetables

Agriculture is dominated by man's need for basic food-stuffs containing energy-giving carbohydrates and body-building proteins. Throughout the world vast quantities of cereals and vegetables are grown, both for direct utilization by man and also for animal feeds. No other group of food plants have been so important to mankind as the cereal grains. Cereals provide nutrition for more than two billion people and their production is in excess of one billion tons annually. Without the cereals, civilization as we know it would not exist.

Fruits

Even after man had settled down into some sort of communal life based around a primitive agricultural system, searching out wild fruits and hunting for game was still an important part of his existence and contributed in no small way to his survival. Until quite recently fruit production remained a rather casual and local affair, but with improvements in freezing, canning, storage and transportation, the fruit industry has increased in size and importance. Now, fruit tends to be grown in large orchards and plantations which require complex equipment for the maintenance, harvesting and processing of the crop.

Botanically, fruits are the matured ovaries of angiosperms, but this distinction is not always clear to the layman. For example, nuts are fruits, or at least very the seeds of fruits, but very often they are placed in a distinct group. The word fruit conjures up the picture of a fleshy, pleasant-tasting and juicy delicacy picked from a tree or bush and it is in this sense that they are considered here. Fruits exhibit a tremendous range in taste and nutritional value, from the watery pumpkin through firm and fleshy products such as apples and peaches to high-carbohydrate-containing dates. Many fruits however, contain little energy, but instead are very watery. Nevertheless many are good sources of minor minerals and vitamins and so form an essential part of our diet. Fruits can be divided into three main groups: fruits from herbaceous annuals; perennial fruits from temperate climates; and tropical fruits.

The first group includes tomatoes, peppers and aubergines (all members of the Solanaceae) and melons and squashes (members of the Cucurbitaceae). Some of these are commonly thought of as vegetables but are botanically true fruits. They are usually grown as summer annuals in temperate climates and so tend to be rather seasonal in availability.

Included in the perennial fruits from temperate climates are the so-called 'stone fruits' such as peaches, apricots, cherries and plums. The fruit is referred to as a drupe and consists of an outer skin (epicarp) containing the thick fleshy part we eat (mesocarp) which surrounds the stone (endocarp) containing a single seed. Apples and pears are also found in this group. They are referred to as pome fruits and are really 'false fruits', the greater part of which is developed from the receptacle of the flower and not the ovary. For example, the part of the apple which we eat represents the receptacle,

Citrus aurantium
Seville Orange

the core being the ovary. Also included are berries, which are many-seeded succulent fruits such as gooseberries and grapes, or aggregate fruits such as blackberries and strawberries. An enormous amount of investigation has been undertaken concerning the propagation, growing and handling of temperate-zone fruits and their economic utilization probably represents one of the most intensively mechanized parts of agriculture today.

A huge range of tropical fruits exists reflecting the wide diversity of tropical vegetation. With the exception of bananas, pineapples and citrus fruits, tropical fruits are not so widely cultivated as are those from temperate zones although some, such as avocadoes, mangoes and guavas are becoming more widely available with more rapid transportation methods. The majority of tropical fruits such as tamarinds, loquats and breadfruit, although widely distributed, are very often utilized only where they grow.

Sugars and starches
Sugars are the soluble forms of carbohydrate in plant cells and can be directly utilized for energy and growth. Starch is an insoluble form which is used for food storage.

Although there are many different sugars which occur naturally, only a few are extracted and used in any quantity, the most important one being sucrose (commonly known just as 'sugar') mainly obtained from sugar cane or sugar beet. Over 60 million tons of sucrose, initially obtained from plants as an expressed juice, are produced annually. Sucrose is used not only as an additive to other foods and drinks to make them more palatable but also for fermentation and as an industrial raw material.

The main growing areas of sugar cane in the Caribbean, South America, southern Asia and the Central Pacific account for something like three-fifths of the world's total production of sucrose. Extraction of sucrose from sugar beet accounts for most of the rest of the world's production of sugar. Sugar beet extraction was developed, mainly in Germany in the late eighteenth and early nineteenth centuries, in order to gain some degree of independence from tropical sources. The industry quickly expanded and today forms an important part of the agricultural productivity in all continents except Africa, the main producers being, northern Europe, USSR and USA.

A less important source of sugar is the gathering of maple syrup from the sugar maple (*Acer saccharum*). This was first accomplished by the American Indians who hacked wounds in the maple trunks and collected the exuded sap in crude wooden or pottery vessels. Today, in the early spring, metal spiles (or spouts) are driven about 5cm (2in) into the trunks and the sap allowed to accumulate in a pail. Controlled evaporation of the collected sap then produces maple syrup, annual production of which amounts to several million litres.

Starch can be obtained from a wide variety of plants. Normally the starch-bearing cells are broken open and the abundant starch grains within floated free or separated by centrifugation. Commercially two sources have been exploited: the temperate *Zea mays* (maize or corn) and the tropical *Manihot esculenta* (manioc, cassava, tapioca, etc) although other sources such as potatoes, wheat, rice, arrowroot and sago are also used. Starches have an enormous number of uses and are encountered in modern economy almost everywhere. They are found in various foods, such as puddings, chewing gum, bakery products, in adhesives and pharmaceuticals, as textile stiffenings and paper coatings and, following hydrolysis, in a whole number of products from sweets to tobacco flavourings.

Beverages
Man has always sought to make his drinks more palatable and more interesting to the taste than is pure water. An obvious alternative to plain water is fruit juice, either used as such, or fermented to make wine or cider. One step removed from this are the diffusions and decoc-

Wine is probably the fermented drink first made by man and today the grapevine is cultivated for its production all over the world in suitable climates. It probably originated somewhere near ancient Armenia (now north-east Turkey).

tions from various processed plant parts, for example tea and coffee, which are so commonly used today.

Tea is obtained from the tea plant, *Camellia sinensis,* and is produced mainly in Southeast Asia. China, India and Sri Lanka being the prominent growers. The tea bush is a flowering shrub which grows best on a well-drained acid soil in a warm climate which has ample rainfall (150–500cm (60–200in) per year). The plants are usually heavily pruned and not allowed to reach more than 1–1·5m (3–5ft) in height. Harvesting is usually done by hand to ensure the best quality tea and only the terminal two leaves (sometimes terminal three or four in poorer quality teas) are plucked.

The processing of the harvested leaves is quite complicated; for the black tea favoured in the West, the leaves have to be withered, rolled, fermented and fired, while for the green tea used mainly in the East a less elaborate procedure of drying and rolling is employed. The quality of the tea produced is dependent on a number of factors: care taken in cultivation, rainfall and elevation at which it was grown. The flavour is dependent on its content of essential oils and other soluble components, while its refreshing and stimulative properties are the result of its caffeine content which occurs in the leaf in concentrations of 2–5 percent.

Coffee is another beverage which depends upon caffeine for its stimulating properties. Coffee is obtained mainly from an Abyssinian shrub called *Coffea arabica.* This plant has glossy deep-green leaves and bears fragrant white flowers two or three times a year. The flowers give rise to two-seeded drupes (the coffee berries), each tree producing something like 3kg (6·6lb) of berries per year.

One of two methods is used for processing following harvest. If water is scarce, the gathered berries are dried in the open and then stored in bins for later processing to obtain the seed following removal of the outer pericarp. In the alternative 'wet' process, which produces better quality coffee, the outer fleshy layers of the berries are removed by pulping machines. After thorough cleaning, the de-pulped fruits are further treated to obtain the familiar coffee beans ready for roasting. The roasting process is essential for the development of flavour, which is due to certain essential oils and caffeol. The process also helps break down the cell walls of the bean and facilitates the grinding process. The stimulant in coffee is caffeine which occurs in the beans to the extent of 1–2 percent.

Cocoa or cacao is made from the seeds (beans) of a small tropical tree originally found and cultivated in southern Mexico and the northern parts of South America. Long before Columbus' visit it was a highly prized beverage in Mexico and Peru; now its cultivation has been vastly extended, with Ghana in Western Africa becoming the greatest centre of production. The cocoa tree (*Theobroma cacao*—'theobroma' meaning 'food of the gods') grows as a many-branched tree with a maximum height of about 8m (26ft). The flowers are borne on the older branches and trunk from which develop large ellipsoid-shaped pods containing between 30 to 50 seeds or beans. The beans are rich in oil, protein and starch but the familiar chocolate taste is not found in the fresh material and must be developed by a process which involves fermentation.

Spices and essential oils

Essential oils in the guise of perfumes and spices have been some of the most sought-after bot-anical products in the world. The search for spices and the resulting intensification of the spice trade influenced the course of history and encouraged the setting up of colonial rule, the repercussions of which we are still experiencing.

The essential oils, responsible for the aroma of spices and the fragrance of flowers, are widely distributed throughout the plant kingdom. They are usually small molecules, typically liquids, possessing an aromatic fragrance which is easily detectable owing to their volatility in contact with air. They may be produced in different parts of the same plant, such as leaves, bark, roots or seeds, and are commonly secreted into special glands from which they can pass through the plant cuticle into the air. Essential oils are most often extracted from the solid portions of the plant tissue before their use in perfume products and food flavourings—they are also used as antiseptics, insecticides, solvents in the paint industry, components of glues, inks, polishes, etc. Most spices consist of the plant parts themselves and are ordinarily prepared by drying and grinding, such processing being carefully undertaken to prevent loss of the essential oils by volatilization.

Cinnamon is a very ancient spice obtained from the bark of *Cinnamomum zeylanicum.* The cinnamon tree is small and bushy with thick shiny leaves and rather inconspicuous flowers. The bark contains about 1 percent of a volatile oil, mainly cinnamic aldehyde, and is either used directly or extracted for the oil. The main production is from Sri Lanka and the Seychelles Islands but trees are also grown in southern India, China, Malaysia and Indonesia.

Clove trees *(Eugenia caryophyllata)* flourish best on well-drained soils in maritime climates with 220–250cm (86–98in) of rainfall annually. The main areas of production today are the Malagasy Republic and Indonesia. The cloves of commerce are the dried unopened flower buds from the trees, and when mature each tree may yield up to 35kg (77lb) of dry cloves per year. The deep rose-pink clusters of buds are harvested and then sun-dried on matting: they are then either marketed as such or distilled to produce clove oil which contains 80–95 percent eugenol.

Ginger comes from the sun-dried rhizomes (root-stock) of a tropical herb, *Zingiber officinale.* The plant is widely grown in tropical countries but most of the world's supply of ginger comes from India and Taiwan. The pieces of harvested rhizomes are prepared for market by peeling, washing and drying in the sun.

Yet another tropical tree species which acts as a source of spices is the nutmeg tree *(Myristica fragrans).* Nutmeg is cultivated principally in the West Indies, Indonesia and Sri Lanka. The tree bears fruits which are like orange–yellow plums. When ripe the fruit splits open to reveal a brilliant red membrane called an aril that surrounds the seed. When dried, this membrane fades to an orange–yellow colour and forms the well-known spice mace. When the seeds are dry they are removed from their shells for use as nutmeg.

Pepper, from the pepper plant *(Piper nigrum),*

Rice, one of the most important cereal crops in the world, is usually cultivated on sloping ground (sometimes very steep) in 'paddy fields' following the contours, the essential water draining down from one level to the next. Here a woman is putting seedling rice plants into previously cultivated ground.

is the most important of all spices in terms of world trade. Peppercorns are the dried fruits of this clinging vine with shiny leaves and adventitious roots which enable it to cling to the trunks of trees. It is mainly produced in India and Indonesia but some also comes from Brazil and the Malagasy Republic. Black pepper is produced from the ripening berries which are picked soon after they turn from green to red. As they are dried the berries turn black, when they are ready for marketing. White pepper is produced by allowing the berries to ripen further and then removing the outer black skin by boiling or fermentation.

Spices from temperate regions perhaps seem more prosaic than those of the tropics, but nevertheless they make an important contribution to food flavourings in many parts of the world. Many of them belong to the family Umbelliferae (such as dill, anise, parsley 'and coriander) while another important group is the Labiatae (such as sage, mint and savory). Many of these herbs are grown on a kitchen-garden basis but limited growing of the more important ones is practised.

Peppermint (*Mentha piperita*) and spearmint (*M. spicata*) began to be important commercial crops in the early twentieth century. Usually the plants are machine-mowed and the mowings allowed to dry in the fields to reduce their moisture content. The crop is then collected and usually distilled to extract the essential oils, the most abundant constituent of peppermint oil being menthol while that from spearmint is carvone. World production is largely centred in the USA, USSR and Japan, where a different variety of mint, *M. arvensis* (Japanese mint) is grown.

Sage *(Salvia officinalis)* is a many-branched plant with slender greyish leaves native to the Mediterranean area and Yugoslavia is the main centre of commercial production. The foliage is harvested when the plants begin to bloom, normally by mowing and raking. The dried leaves contain less than 2 percent of various essential oils and are used almost exclusively for food flavouring.

Household uses for plants
Wood

For a long time wood was the principal substance used by man for building his home and provided the means to furnish it as well as the utensils to use in it. Even in this atomic age, wood is still a vitally important commodity commanding high prices and requiring large resources of space and manpower for its production. Wood is very often used directly, the nature of use depending on the type of wood itself, but the use of 'processed' wood and wood products is discussed here.

The paper used for newspapers and books is very often manufactured from wood pulp and most probably the paper on which this book is printed was once part of a forest in Scandinavia, Canada or USA. About 80 million tons of pulp are produced annually throughout the world, half to make paper other than newsprint, one-third for cardboard and about one-sixth for newsprint. Pulp is produced mainly from softwoods or conifers: two qualities of pulp are obtained, one by mechanical means (grinding) and the other by chemical means (dissolution of binding materials such as lignin to free the cellulose fibres), the latter process giving the best quality paper.

Above: Mechanical harvesting of sugar cane, a very important tropical crop. The stems of the plant are repeatedly pulped between rollers, and the extracted juice is evaporated and refined. The fibrous residue is used as fuel and, increasingly, in making fibreboard and paper.

Top right: From the pods of Vanilla, widespread tropical climbing orchids, the flavour of the same name is extracted. It is used to flavour food and tobacco and in perfumery. Vanilla is one of the few useful products obtained from the vast orchid family; today most is made synthetically.

Right: Logging is always a difficult operation because of forest terrain, needing the use of winches, ropeways, trucks, trains and even elephants. Here logs are being hauled from an Australian eucalyptus forest. Eucalyptus, which is being widely planted in many parts of the world, is a good hardwood.

The leather industry depends upon the successful processing of animal hides to produce a tough stable leather, resistant to oxidation, moisture, extreme temperature changes and bacterial attack. An important part of this process is tanning, in which tannins (a complex group of astringent substances found almost universally in plant tissues) are allowed to combine with the proteins in the hides over a period of time which can range from a few hours to several months. Although tannins can be ob-

tained from numerous sources, they are most often produced from tree bark by leaching with hot water. The tannin concentrate so produced is then used directly for the tanning process. Oak, chestnut and hemlock trees have been traditional sources of bark for tannin extraction: present-day commercial sources of importance are wattle and mangrove barks and chips of quebracho wood.

Although not directly obtained from wood, rubber can perhaps be considered as a 'wood product'. Rubber is manufactured from the latex (a milky juice) present in the cells of several species of tropical and subtropical plants. Most important is para rubber obtained from *Hevea brasiliensis,* a tree first found in the forests of the Amazon and transferred in the nineteenth century to Sri Lanka and Malaysia where eventually over 90 percent of the world's supply of rubber came to be, and still is, produced. Normally, the latex is obtained by tapping the tree trunk, using variously shaped incisions, and collecting the exuded latex in specially designed cups. After collecting, the latex is coagulated, either by acid or by smoking, and then sold for processing to make the wide variety of articles that are so familiar to us today. Other plants, including dandelions also produce latex.

Other 'household' plants

Wood is not the only plant product of material use to the household. Plant products are used to clothe us, to colour our clothes and to perfume our bodies; they provide us with ropes for tying and binding and materials for basket weaving. Although many of these uses have been superseded by the discovery of synthetic substitutes there is nevertheless a sense of nostalgia in many people which encourages the continuation of traditional recipes and methods. Of these, the use of plant extracts for colouring material is perhaps one of the most interesting aspects.

The employment of colour must be as ancient as man himself and for this purpose he has always made extensive use of vegetable dyes. From their rather primitive use on parts of the body, as ornamentation or as protection against evil spirits, their use was developed extensively for dyeing clothes, a method which continues today, although most dyes are now synthetically manufactured.

In western Europe one of the earliest products was woad, a blue dye obtained from *Isatis tinctoria* involving a very smelly fermentation process. Combination of woad with other natural dyes such as madder powder (red) or weld (yellow) produced a wide range of other colours so that in the Middle Ages woad was referred to as 'universal dye'. Eventually woad was supplanted by indigo, another blue dye, obtained from the plant *Indigofera tinctoria* (actually containing the same blue pigment, now called indigotine) and which had been used in Asia for over 4,000 years.

Of the yellow dyes one of the most sought after was saffron, the principal yellow colouring used by the Greeks and Romans. Saffron is obtained from the yellow stigmas of a blue crocus *(Crocus sativus)* which is cultivated mainly in the Mediterranean area. Because it is only obtained in relatively small quantities (about 8,000 flowers are needed for the production of 100g (3·5oz) of dried saffron) the use of saffron as a dye has been replaced by cheaper substitutes, but it is still valued as a food additive.

For producing a red colour one of the most important dyes was obtained from the madder plant *(Rubia tinctorum)*. The red pigment can be extracted from the root-stock and may have been first used in India although it was well-known to the Persians and ancient Egyptians. With the advent of modern synthetic dyes the use of madder for dyeing has been superseded although it is still produced in small quantities for artists' needs (e.g. Rose Madder and Madder Brown).

Plants and health

The use of various herbal mixtures and in[...] for curing all manner of illnesses has been [...] tised from time immemorial. Drugs cap[...] causing death or curing diseases were [...] considerable awe by the layman and t[...] gether with the sometimes elaborate [...] which accompanied their dispensation [...] often have rendered simple herbal r[...] much more efficacious through the [...] logical effect upon the sick person. N[...]

A bumper issue of the New York Times [...]
roughly 400 hectares (1000 acres) of conifero[...]
like this—no wonder that cutting timber is [...]
industry. More obvious uses for timber are f[...]
ings, boats, tools, furniture, and weapons, an[...]
products like bark, resin and latex have furth[...]

less, some of the herbal remedies used really were effective because they contained specific substances which had physiological activity. In many cases studies of these herbs has enabled the active principle to be identified and has occasionally led to the production of whole new ranges of compounds.

In addition to the use of pharmacologically active herbal preparations, many plants contribute to the general health of man by virtue of their vitamin content. Whole books have been written on the nutritional value of foods from this standpoint and it is worth noting that a large proportion of our daily vitamin requirements is very often derived from plant sources.

Painkillers

One of the earliest and most famous (or infamous) of the painkillers was the root of mandrake *(Mandragora officinarum)*. The fleshy root of this plant is very often forked and sometimes bears a rough resemblance to a human figure, hence in mediaeval times an air of mystery and legend surrounded its use. Steeping or boiling the root in wine produced a concoction which when taken in the proper dose induced sleep and insensibility to pain. Too much however could cause paralysis, madness or death (the preparation came to be known as 'death wine' or morion). The active principle in mandrake has been shown to be the alkaloid hyoscine which also occurs in other plants such as the corkwood of Australia *(Duboisia myoporoides)* and the Indian thorn apple.

Hyoscine is still used in modern medical practice, sometimes as an anaesthetic in combination with morphine, which is obtained from the opium poppy *(Papaver somniferum)*. The opium poppy is cultivated in many parts of the world, especially India, China and in the Near East and Mediterranean area. Opium is the dried milky juice or latex obtained from the poppy seed capsules which are cut some two weeks after the petals have dropped. The latex exudes from the cuts and in hot dry weather congeals into lumps which are subsequently processed for their morphine content (the latex can contain up to 15 percent of morphine together with 1 percent of codeine and various other alkaloids). Morphine is widely used as a painkiller although the abuse of the parent opium and related heroin (acetylated morphine) as narcotic drugs is one of the great social problems of today.

Another well-known anaesthetic is cocaine, the active principle of the coca plant *(Erythroxylon coca)*. The coca plant is a small tree or shrub with ovate leaves from which the cocaine [...] [ex]tracted. Although synthetic substi[...] for cocaine (for example, novo[...]ine) the coca plant has become [...] [s]timulant used by Peruvian [...] ertaking long arduous jour[...] that they can work for days [...]od if they chew the leaves of [...]

[...] drugs

[...]y narcotics and drugs obtainable [...] ariety of plants. One of the most [...] quinine which is obtained from [...]cies of *Cinchona* trees. *Cinchona* is [...]he jungles of Peru and following the [...]y of the antimalarial properties of the [...]ed bark, plantations were eventually [...]hed (not without various degrees of [...]e), in India and Java. Four alkaloids are [...] in *Cinchona* bark, all of which have anti[...]rial activity. These are cinchonine, cin[...]nidine, quinine and quinidine, with some species of tree yielding a total of 17 percent of them in their bark.

Many more drugs come from the tropical forests of South America. Perhaps the most well-known of these (or most notorious) is curare, a complex resinous mixture prepared by boiling together an assortment of bark and herbs and finally evaporating the extract until a thick black residue remains. The poisonous mixture is used for tipping blowgun darts and arrows to kill animals and birds. Many plants usually go into a brew and maybe no two potions are exactly identical. *Chondrodendron tomentosum* is very often an ingredient together with plants of the *Strychnos* species. *C. tomentosum* provides the active principle of the curare which has been shown to be an alkaloid named d-tubocurarine. The toxic properties of *Strychnos* species are due to their contents of two alkaloids, strychnine and brucine. When used in the proper dosage these alkaloids can relieve paralysis and stimulate the central nervous system, but in excess have the opposite effect and in large doses cause death.

Not all drugs are obtained from tropical or subtropical plants. One very useful drug is digitalin obtained from the foxglove, *Digitalis purpurea,* a common European wildflower and grown as a garden plant all over the world. Digitalin containing extracts are prepared from the dried pulverized leaves of the plants, yields of up to 300kg (660lb) per acre being possible. The most important medical uses of digitalin are as a heart stimulant and diuretic.

The search for useful drugs from plants is still continuing and is likely to continue for many years to come. Recent research has shown that an alkaloid extracted from the flowers and developing fruit of *Camptotheca acuminata* (camptothecin) has anti-tumour activity in tests using laboratory animals. Maybe this plant or a similar one will provide man with the long-sought-after cure for cancer, but even if this is not so there is no doubt that plants from all over the world will still be employed and investigated as possible sources of drugs beneficial to man.

Hallucinogens

No account of drugs would be complete with-out mentioning the hallucinogenic or psyche-delic drugs so familiar to us in the twentieth century. Mescaline, from the cactus *Lophophora williamsii,* pantherine from the fungus *Amanita muscaria* and psilocybin from another fungus, *Psilocybe mexicana* have been used for centuries by primitive peoples. The two best known hallucinogens today are LSD (lysergic acid diethylamide) and marijuana (tetra-hydrocannabinol). LSD is easily synthesized but was originally found in plants of the Convolvulaceae, particularly *Ipomoea violacea.* Apart from the social problems arising from its use, it does have some promise as a tool in the treatment of schizophrenia. Marijuana is an ancient drug, described as early as 2737 BC, and is obtained from the hemp plant *Cannabis sativa.* Its use has led to considerable discussion and argument concerning the social problems and possible addictive properties of such drugs, with currently little agreement between the various authorities.

Insecticides

There are innumerable plant products which can have intense physiological effects upon man. It is not surprising therefore that there are some plant products which can affect the physiology of pests such as insects even to the extent of destroying them. For the most part synthetic insecticides overshadow the use of natural products but in certain situations the use of the latter is preferred. The best known natural insecticides are rotenone, obtained from *Derris* species and *Lonchocarpus* species, and pyrethrum, obtained from the daisy-like flowers of certain species of *Chrysanthemum.*

Right: Indian hemp, **Cannabis sativa,** *an annual relation of the mulberry, was first grown in China 4,500 years ago as a fibre plant; but today it is cultivated almost entirely for the drug marijuana or hashish.*

Below: As its name suggests, the Death Cap, **Amanita phalloides,** *is a totally lethal fungus if eaten. Yet its relation A. muscaria yields a hallucinogenic drug, and other relations are esteemed as food.*

CONSERVATION
endangered plant life

The plant kingdom's great richness is in severe danger of becoming drastically degraded as a result of man's activities. At least 25,000 species, comprising around ten percent of the world's flora, are believed to be threatened with possible extinction or to be dangerously rare. This figure is all the more serious when one realizes that man depends totally on the plant kingdom for his basic needs of food and oxygen, as do all other animals, and that plants provide timber, numerous drugs and a host of other incredibly varied products. In addition he delights in the charms and attractions of numerous widely differing species, both in the wild and in the garden.

There is a terrible finality about the extinction of a species. As with the dodo of Mauritius, once gone it is lost for ever and its unique set of characters cannot be reconstituted. With plants the situation is all the more tragic since many species may become extinct before their possible value is known; the economic potential of so few plants has been investigated and man has made use of only a minute fraction of the plant kingdom. Modern research is constantly discovering new products in obscure and little known plants. For instance, the seeds of *Simmondsia chinensis,* a shrub known as jojoba, contain an oil recently found to be a substitute for the precious oil only previously known from the endangered sperm whale. It may well be of great economic value in the future. Many of the estimated 25,000 threatened species may have similar potential; their loss would be an inestimable tragedy.

The best way of looking at the threats to plants is to consider a few examples of threatened floras, in particular three that are probably the most in danger: those of tropical rain forests, of islands and of the countries of the arid zone. These are only a selection of threatened habitats and areas but it may help to stir concern for man's most precious resource—the cornucopia of the plant kingdom.

Tropical rain forests

The tropical rain forests of the world are becoming seriously threatened and there is concern over their future. These evergreen forests occur in areas with a tropical climate with rain throughout the year and little or no seasonality. They are the principal vegetation in coastal West Africa and in the Congo basin, in Southeast Asia from the Malay Peninsula through the islands of Sumatra, Java and Borneo to New Guinea and the Philippines, and in South America reaching from the foothills of the Andes across the vast Amazon basin up to the Caribbean and parts of Central America.

These forests are perhaps the most luxuriant and magnificent vegetation on earth, teeming with plant and animal species in an endlessly changing continuum and are especially interesting for the botanist for their small isolated pockets of very local species. At the other end of the scale, many rain forest species have unusually scattered distributions, occurring in large overlapping areas at a density of only a few individuals per square kilometre.

The forests are particularly rich in climbers, palms and epiphytes—in particular orchids and bromeliads in South America and orchids in Southeast Asia. There are believed to be 3,000–4,000 species of orchids alone in the area from the mainland of Malaya to New Guinea. This is a remarkable figure—more than twice the whole flora of the British Isles.

Rain forests are also of great significance since, with the exception of areas too inhospitable for human settlement, they are the only type of vegetation of which large tracts remain,

or did until very recently, untouched by man. Evolution in some rain forest areas may have continued uninterrupted under favourable climatic conditions since the flowering plants first evolved. Despite the immense problems in allocating land use to provide food for fast-expanding human populations, it is vital to preserve at the very least large intact samples of such forests so that the processes of evolution can continue.

The insatiable world hunger for wood is a major threat to the rain forests, especially in the countries where the finest timber trees, especially the Dipterocarpaceae, are found. In West Malaysia, at current rates of extraction, present trends indicate that all the lowland dipterocarp forest outside forest reserves and pro-tected areas will have been exhausted within the next ten years. For example, it has been estimated that 2 sq km (0.8 sq mile) of lowland dipterocarp forest per day passes over the cause-way from Malaya into Singapore for export. Another example is of 172,000 hectares (425,000 acres) in the Philippines lost every year. Further threat comes from new technology to remove not only the logs but virtually the whole vege-tation for chipboard. Unlike a temperate forest, most of the nutrients in a rain forest are tied up in the vegetation rather than the soil and so removal of the plant cover can result in loss of soil fertility, especially in upland areas.

In the Amazonian rain forests the principal threat to the flora is the uncontrolled utilization of the forest following the opening up of the area for development. In an effort to expand the potential of the region, huge road-building projects are underway, such as the 5,619km (3,510 miles) Transamazônica Highway from the Atlantic to Peru. The scale is immense and the aim is to use much of this vast area for commercial exploitation in terms of minerals and agriculture. On both sides of the roads, vast estates are being cleared for cattle-ranching. However, little of the Amazon soil is likely to be suitable for agriculture; within a few years

This Turkish hill has been denuded of its trees by man seeking timber for building and fires, and by the effects of grazing animals. Topsoil has been blown and washed away and deep eroded gullies have formed.

Above: Man has cut down almost all the trees of this Colombian rain forest in order to raise crops and pasture cattle. Only a few specimens of the palm Syagrus sancona— an endangered species—remain for the present.

Left: Untouched tropical rain forest in the Merangin gorge, central Sumatra. Lush and dense, indeed virtually impenetrable, it is a well integrated plant community containing many different species, from shade-lovers to tree-top epiphytes.

the quality of the grazing will decline rapidly. On plots cleared for food-crops, the decline in soil fertility, the increasing weeds and the burgeoning insect pests, all associated with peasant agriculture, are tending to lead to degradation of the land. Ignorance and lack of fertilizers are crucial factors.

The flora of the Amazon region is probably the least explored anywhere in the world and numerous species are doubtless being lost before they can be discovered. Rain forests in South America, when severely disturbed on a large scale, may not regenerate. When exploited in this way, the forest becomes a non-renewable resource like coal or oil, but one in which the diversity of its renewable components is being inexorably lost.

There are abundant reasons why conservation is necessary. Strong arguments centre round the value of the upland forest in maintaining water catchment and preventing soil erosion. The production of hardwood timber is also important since there will always be a demand for the quality hardwoods despite the great pressure for paper, chipboard and similar products. But the foremost argument is that the rain forests, in all their luxuriance and diversity, represent a gene-pool of inestimable value to man.

The potentials of the great majority of the plants have never been explored and many could well prove to be of vital importance in the future, either as crop-plants in their own right or as material for use in crop-breeding. For example, the Southeast Asian forests are particularly rich in trees with edible fruits, but

only two species of the genus *Durio* are in cultivation out of 19 species in Borneo, most of which have delicious fruits. No known work on breeding has been done. Of the mangoes, only *Mangifera indica* is widely cultivated and bred. Other species in the genus have only been grown very locally, although some are excellent, such as one from a small area of Borneo which has fruits as large as a coconut. Similar situations exist for plants producing oils, drugs and all manner of useful products. In addition there are numerous species of interest to botanist and visitor, like the giant *Rafflesia arnoldii,* an extraordinary parasite with the largest flowers in the plant kingdom which is now becoming very rare because of habitat destruction. However the loss of the forest is not only the loss of a uniquely valuable set of organisms—it is also the loss of a whole community, in this case the most prestigious and exciting ecosystem on this planet; composed not only of plants but of animals of every kind which cannot exist outside it.

Island floras

Island floras, especially in the tropics and subtropics, are in particular hazard. The plants whose survival is threatened are mostly the endemics (i.e. confined to the island concerned). These endemic species may have evolved on the island or may be old relicts of formerly widespread species which have died out elsewhere. In either case, because of their isolation from outside evolutionary pressures, they are particularly susceptible to grazing from introduced animals and competition from vigorous introduced plants.

Consider the flora of Socotra, a dry island 225km (140 miles) east of the Horn of Africa, Cape Guardafui in Somalia. It has been known since antiquity as a source of dragon's blood, a red dye produced from *Dracaena* species. From a total of about 600 species of flowering plants, 216 are at present believed to be endemic. As a result of grazing by excessive numbers of livestock, at least 85 of these are in immediate danger of extinction. The flora has presumably evolved partly in the absence of large mammals, unlike the flora of the African mainland, and few species have characters such as spines which act as a defence against grazing animals. Hence they are particularly susceptible to the ubiquitous goats. In 1880 Professor I. B. Balfour on a botanical expedition to Socotra found the prickly *Lasiocorys spiculifolia* to be 'one of the plants which makes progress over many parts of the plains unpleasant'. However, in 1967 the vegetation on the coastal plains had become so degraded that only a single individual of this species was found on the whole island.

In many cases the species concerned are of great importance or interest to man. The only species in the pomegranate family other than *Punica granatum* (the cultivated pomegranate) is a small tree endemic to Socotra called *Punica protopunica.* In 1967 only four very old and widely separated trees were seen and when these trees die, it will probably become extinct in the wild. The goats prevent any regeneration by eating any young seedlings that may come up. Near relatives of crop plants are extremely important to man since they may be needed for breeding factors such as resistance to pests and diseases into the cultivated varieties. Another endangered Socotran species is *Dirachma socotrana* whose flowers are so distinct and unusual

that the species has been given a family of its own, the Dirachmaceae. In 1967 it was reduced to a grove of about 30 trees with no regeneration and unfortunately it is not in cultivation. Its extinction would represent a major loss of diversity in the plant kingdom.

The island of St Helena in the South Atlantic is another example showing the catastrophic effects of introduced plants as well as grazing animals. Goats were introduced in 1513 and within 75 years there were sightings of herds 2km (1.2 miles) long on this small island. They devastated the native forests. The island was explored botanically for the first time in 1805–10 when the forests had already been reduced to a few patches, high up on the central ridge. About 31 endemic species are known, of which 11 are extinct. However the great Victorian botanist, J. D. Hooker of Kew, estimated that there must have been about 100 endemic species in this 'wonderfully curious little flora' before man intervened. These plants will never be known.

The goats have now been controlled and the principal threat is introduced plants such as New Zealand flax, *Phormium tenax*, and gorse, *Ulex europaeus*. These are spreading rampantly through the small patches of relict forest. Since the flax industry declined, the crop has not been cut and so the plants have seeded abundantly. Since flax is about 3m (10ft) tall, it swamps small endemics such as *Wahlenbergia linifolia,* a low shrub with delightful white bell-flowers close to *Campanula*.

Similar scenarios can be spelt out for numerous islands, particularly those in the Indian and Pacific Oceans. Mauritius, the Seychelles, Sri Lanka, the Marquesas Islands and the Juan Fernández islands may sound like tropical idylls, but similar botanical tragedies have happened there. The Juan Fernández Islands, the home of Robinson Crusoe, lying off the coast of Chile, are particularly badly damaged due to introduced sheep, cattle, horses, goats, rabbits, rats, and plants such as *Aristotelia* and *Rubus*

(bramble). On many islands goats and other mammals were introduced by mariners several hundred years ago so that there would always be fresh meat available for passing ships. The effect on the plants has been catastrophic.

Pressures on land for tourist developments, for agriculture and for forestry are of course particularly acute on islands because of the problem of size. In the Canaries, with about 500 endemic species of flowering plants, wide-scale clearance of hill land for agriculture is having a disastrous effect on the flora. The Canaries are the home of a very special vegetation, the laurel forests, which have many endemics. They are the only areas on the island which remain moist throughout the dry Mediterranean summers and they play an important role in condensing moisture from the clouds as well as preventing erosion on steep slopes. Removal of the laurel forests and excessive coppicing of the trees for poles and firewood results in erosion damage and the

streams lower down the hillsides dry up in the summer. The destruction of this community has now gone so far that even the constituent trees such as *Arbutus canariensis* are threatened plants.

One of the saddest cases of all is Hawaii. It is estimated that 96.1 percent of the plant species that occur on the Hawaiian Islands are endemic. This is an extraordinarily high percentage, reflecting the situation of the islands as a 'laboratory of evolution'. Two hundred and seventy-three species are listed as extinct and 800 as endangered—and the list is by no means complete.

The islands are volcanic and many species tend to be restricted to a few islands of vegetation called kipukas, surrounded by lava flows. Many have become greatly depleted by lava flows in the geological past, and have entered the era of man already very rare and hence especially vulnerable to the grazing and habitat destruction which followed. For example,

Hibiscadelphus has flowers like those of a *Hibiscus*, but by failing to open completely are adapted for pollination by endemic birds of the nectar-feeding *Drepaniidae*, the Hawaiian honey-creepers, with bills fitting the slight curvature of the corollas. It is thus a fascinating example of closely linked evolution between plant and animal endemics. However all the *Hibiscadelphus* species and most of the honey-creepers are either endangered or extinct and it is uncertain whether one can survive for long without the other.

Arid zones

In many countries of the arid and semi-arid zones, overgrazing by domestic livestock is having a catastrophic effect on the vegetation, in places reducing pasture to desert and semi-desert as, for example, has happened in the belt from Western Asia to the Sahel in Africa.

The effects of overgrazing on vegetation are

Above: The endemic giant echiums of the Canary Islands, like E. pininana *which reaches 4m (13ft), are becoming very rare, owing partly to the destruction of the laurel forests by farmers. Only a small population remains.*

Above left: The goat, probably man's earliest domesticated herd-animal, is tough, adaptable, multiplies rapidly and is capable of eating almost anything green. Its long-term effect on vegetation is usually disastrous, especially as it eats tree seedlings.

complex. Since animals are selective in their feeding habits, the composition of the flora gradually changes. Goats, sheep and cattle all have somewhat different effects and so the proportion of the various animals in the herd is significant. Goats in particular eat virtually anything and so prevent regeneration of nearly all the flora. However, despite the differences, a common pattern often emerges. Firstly the palatable herbs tend to become eaten off and the grasses reduced. The trees cannot regenerate and cutting for firewood and timber may accelerate their decline. At this stage the soil may begin to blow. Eventually even the few bushes that are left and even the unpalatable plants such as aloes may disappear as the land becomes a desert or semi-desert.

For example, on the Red Sea Hills of Ethiopia and the Sudan, the distinctive palm-like tree, *Dracaena ombet* of the Agave family, is now drastically reduced. It was formerly a main component of the scrub covering these extensive hills, but all the plant life over vast areas has been eliminated and only occasional and scattered trees remain, surrounded by bare rock. The ombet's decline was hastened by the removal of the trunks for firewood and the leaves for making mats and baskets.

In most cases it is not yet known which plants are threatened. Many countries such as Turkey and Iran have particularly rich floras. Elsewhere the floras are not so well known and few have been looked at in terms of conservation, but in 1972–3 an expedition to Somalia

Right: The Lady's Slipper Orchid, Cypripedium calceolus, is on the road to extinction because of its beauty. Now only known in one British locality, it is becoming very rare throughout the north temperate zone it inhabits.

Below: The main locality for Phoenix theophrasti, the very rare Cretan date palm, is a grove in a river valley leading to a fine sandy beach, where tourists are beginning to damage this unique plant.

aimed to do just this. Many species were found to be on the verge of extinction, especially the endemics which are believed to make up as much as 30 percent of the flora. The results of overgrazing were very extensive, especially in the north where visitors in the nineteenth century had found a country with park-like vegetation of scattered trees and abundant grass, typical of much of East Africa. There was much game, a valuable source of meat.

Today this same area is one of the most degraded and few trees are left. The last elephant was said to have died in 1953. Large areas are now dominated by species such as *Aloe megalacantha* which is unpalatable to stock and marks one of the last stages in the formation of deserts and semi-deserts. Indeed the whole country has been overgrazed to some extent and the carrying capacity of the land greatly reduced. The main wealth of the Somalis lies in their grazing animals so the situation for the country is very serious. The political and human problems caused are overwhelmingly difficult, but it is clear that the present overstocking is a course for disaster. The vegetation of the arid and semi-arid zones is one of the most fragile ecosystems in the world and is very susceptible to permanent damage from overgrazing.

Threatened European species

In contrast the European flora is not so seriously affected as most of the tropical floras. Nevertheless there are about 1,000–1,200 species in Europe that are threatened or very rare. Many are cliff or mountain species which are only known from a few localities, often fairly inaccessible and so not under immediate threat.

The main risk comes from unscrupulous collectors in many cases, and it is felt that some form of protection and monitoring is desirable to safeguard their futures. The devastation of the vegetation of so much of the Mediterranean coasts by developments for tourism has not had too serious an effect on the flora so far since most of the coastal species are widespread, often occurring both in North Africa and Europe. A rather unusual exception is *Phoenix theophrasti*, the Cretan date palm, which is confined to one river valley leading down to the sea in Crete and a few other scattered localities on the island. As the only palm grove in Europe, the site is a major tourist attraction and the palms are beginning to suffer.

In northern Europe the main group of species that are threatened on a continental scale are the aquatic and wetland plants such as the spectacular insectivorous plant, *Aldrovanda vesiculosa*. Drainage, pollution and infilling have taken their toll on these vulnerable aquatic habitats. In contrast most of the orchids, which are generally thought of as becoming very rare as indeed they are in Britain, are frequent in many countries of southern Europe. The exception is of course that most handsome of orchids, *Cypridedium calceolus*, the lady's slipper orchid, which has been over-picked and uprooted for planting in gardens. Although it grows from Europe through Russia and China to the USA it has become very rare, at least in most of Europe. In Britain only one zealously guarded clump remains.

Protection of threatened species

On a world scale, some collectors are a major threat to certain groups of plants such as tropical

Profile on a threatened species

This delightful small fan palm (Max-burretia rupicola) is confined to three limestone hill tops, all within 40km (25m) of Kuala Lumpur, the capital of West Malaysia. Although two of these hill tops are within the Templer National Park, the palm is at risk on one of them from fires lit by climbers reaching the summit; its total world population is probably below 1,000. The third area is the top of the Batu Caves Hill where it is threatened by quarrying. The Batu Caves themselves are a major tourist attraction and a site of great importance to the Hindu religion. These limestone hilltops, rising as they do above the lowland rain forest of Malaya, are the relics of an ancient calcareous mantle and have an extremely varied flora with numerous orchids; many of the species are very local endemics and hence at great risk from habitat destruction. This species is typical of many—not at present in immediate danger of total extinction but extremely rare and under threat over part of its range.

orchids and succulents which are keenly grown by many amateurs. In Mexico some cacti have been brought to the verge of extinction by commercial raiders. The export of orchids from Southeast Asia is also believed to be considerable. When such a plant becomes rare by natural causes or by habitat destruction, the threat from collectors intensifies since rare plants are considered particularly desirable and fetch high prices. The result is a vicious circle which results in extinction.

A number of countries have strong laws protecting their flora but enforcement is difficult, especially in the tropical countries with such rich and varied floras. Fortunately international action is being taken which may ease the situation. In 1973, 57 nations signed the Washington Convention on International Trade in Endangered Species of Wild Fauna and Flora. By a licensing scheme this aims to monitor, and in some cases control, the international trade in a given list of plants and animals. There are of course facilities to allow legitimate transfer of plants between scientific institutions and to encourage cultivation in large numbers by nurserymen so that the pressure can be taken off wild populations. This is a great step forward.

What can be done about the great majority of threatened plants, those threatened by grazing, forest destruction or similar causes? The best solution is to protect the habitat and so conserve the plants and animals it contains. Cultivation in botanic gardens is a valuable 'long-stop' against final extinction but it is no long-term solution. The formation of National Parks and Nature Reserves to cover large samples of all types of vegetation and to include as many threatened species as possible is an urgent task. The resulting profits from tourism can be considerable, although of course it is very difficult to maintain large wilderness areas intact and free from human settlement, especially where poverty is involved; management must always be related to the needs of the local people. Nevertheless where only small fragments of unique vegetations remain, absolute protection is essential, although it is uncertain how far islands of vegetation in a sea of disturbed land can maintain themselves in the long-term.

In recent years the importance of reconciling conservation with development has received great interest. In Britain this has involved, for example, re-creating habitats for wildlife on small patches of wasteland, leaving aside small areas of trees from the plough, allowing undergrowth to develop in commercial forests. Above all it implies an awareness of the wildlife around us and an understanding of their needs. But it is in the less developed tropical countries where it is so important to bring conservation and development nearer together.

There are very strong economic arguments for conserving vegetation and not just as small isolated areas. In theory this implies not protecting the resource *from* man, but using it on a fully sustainable basis at a level of productivity where none of the various components are lost. It is clear that many areas of over-exploited tropical rain forest, especially in upland areas, will not regenerate. Productivity may fall and the damage may prove irreparable in some areas. It is clear that in arid countries overgrazing inexorably leads towards deserts and semi-deserts. Sadly, most of the countries in this critical position are the ones for whom dependence on the land is total and so failure to observe conservation may spell disaster.

It is for this reason that a major portion of the programme of the International Union for Conservation of Nature and Natural Resources (IUCN) on rain forests and arid lands has been to produce guidelines for development in collaboration with the countries concerned. At the same time, the recently formed IUCN Threatened Plants Committee is steadily building up information on which species are threatened throughout the world and where they grow, so that at least the information needed by decision-makers will be available. Time is very short and the need for reserves and protected areas desperately urgent. Some species now threatened will undoubtedly be lost in the near future

Above: Ultimate desolation . . . There were large trees at Las Anod, Somalia, in 1953, which provided wood, shade, and shelter for smaller plants. Grazing animals and the cutting of the trees have reduced this formerly productive spot to a near-desert. The basic factor behind such man-created wilderness is too many grazing animals creating pressures on natural resources which cannot be withstood.

Left: By eating the young seedlings and reducing the herbaceous vegetation to a tightly cropped sward, grazing animals have prevented all regeneration in this fine old grove of Betula jacquemontii *in Kashmir. This species is perhaps the most spectacular of the white-barked birches.*

and all the time large areas of vegetation are being permanently ruined.

Aldo Leopard, the distinguished American conservationist, is reputed to have once said: 'The first rule of intelligent tampering is to save all the pieces.' This is perhaps the best justification for conservation of plants. It is of the very greatest importance to ensure that the renewable components of the world's vegetation on which man depends, are conserved in perpetuity so as to be available for his future benefit.

HOW PLANTS WORK AND GROW
simple green chemistry

All life depends upon the successful trapping and subsequent utilization of the radiant energy from the sun. Green plants are an essential component of this process because they are the primary energy receptors on which all other life depends. The method whereby plants are able to trap and use light energy is called photosynthesis. Light energy is utilized to synthesize energy-storing carbohydrate from carbon dioxide and water vapour in the air. The reaction is represented as by the following diagrammatic equation:

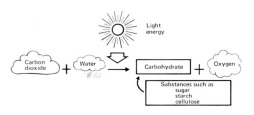

Plants contain special structures within their cells called chloroplasts which control and contain the photosynthetic reaction taking place.

There are two parts to the overall process. First the light energy has to be 'trapped' by the chloroplast and second the trapped light energy is used to 'drive' the synthesis of carbohydrate.

Trapping light energy

Green plants appear green because they reflect green light and absorb the other colours that make up 'white' light. The pigment which gives plants this green colour is called chlorophyll and it is mainly this substance which is used to trap light. The chlorophyll molecules

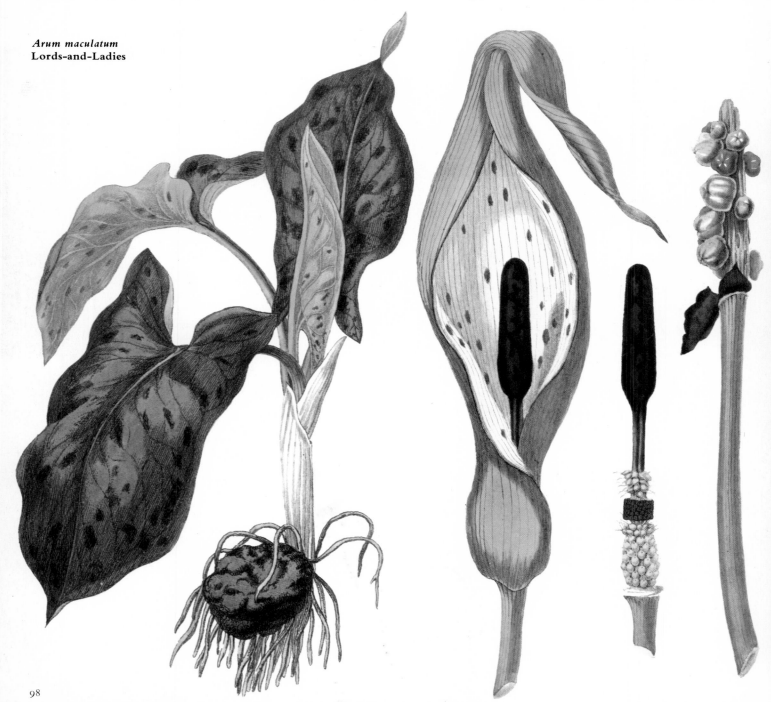

Arum maculatum
Lords-and-Ladies

are arranged in layers within the chloroplast where they act as a receiving array for the light energy.

After the light energy is trapped by chlorophyll it is stored within the plant in the form of 'high energy substances'. These substances are used for transferring energy from one place to another in the chloroplast and they 'drive' the synthetic reactions which make carbohydrates from carbon dioxide and water. During the process of trapping light energy, oxygen is 'split' from water and released into the atmosphere, eg in water plants bubbles of oxygen may temporarily accumulate on the leaf surfaces. (See photograph overleaf.)

Above: Sunlight on beech leaves: all life on earth depends upon the trapping of light energy by green plants.

*Right: Light energy is trapped and used in special structures called chloroplasts. Here we can see these structures in the cells of a moss (**Mnium** sp.).*

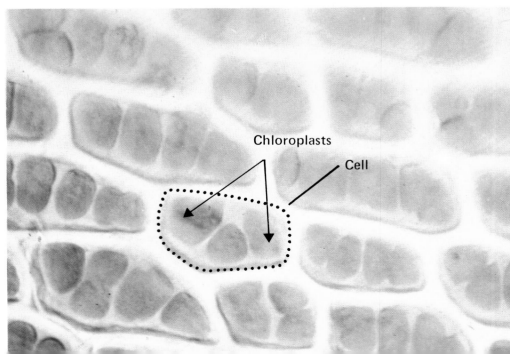

Chloroplasts

Cell

into the mitochondria where they are broken down further into carbon dioxide. During this process the chemical bond energy originally held in the glucose molecule is released and trapped in the form of 'high-energy substances' ready for use at other sites in the plant cell.

The efficiency of respiration in producing usable energy

It has been calculated that respiration is about 40 percent efficient in terms of conversion of stored energy into usable energy. This is a relatively high degree of efficiency as compared with energy conversion in general. The unconverted energy appears as heat, which is used in animals to maintain a stable body temperature, but it is wasted in plants, being dissipated to the surroundings.

The cycle of photosynthesis and respiration

Photosynthesis and respiration represent two processes, one of which is the reverse of the other. Consider an overall cycle in which these processes take place, the purpose of which is to enable the plant to trap the energy of the sunlight, store it and then release it in a usable form for growth.

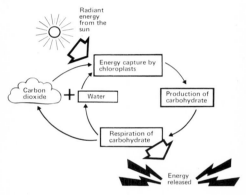

Primary production—the key to all life

All the energy which a plant succeeds in trapping from the sun is ultimately used for its growth and survival. Basically growth depends on the production of all of the molecules required for processes such as building up the plant structure, maintaining the structure once built and production of seeds or other mechanisms for the propagation of the species. Plants, given the abundance of energy from the sun, are able to produce all of their needs from quite simple mineral requirements—man cannot produce all he needs and thus depends on plants for survival.

The most fundamental process involving growth is cell division. In plants the cells which give rise to all other cells are called meristematic and exist only in the tips of roots and stems, in very young leaves and where there is secondary growth such as in trees.

The basis of the cell division process depends upon a duplication of the hereditary material

The synthesis of carbohydrate

The formation of carbohydrate involves the reduction of carbon dioxide gas which is an oxidized form of carbon. The energy needed for this process comes from the 'high energy substances' produced in the chloroplasts. The synthesis of carbohydrates requires a complex sequence of actions which takes place within the chloroplasts, but the products, such as sugar and starch, are transported and used elsewhere in the plants.

The efficiency of photosynthesis in energy conversion

Since all of life finally depends on the conversion of light energy to usable chemical energy it is worthwhile to estimate the actual energy conversion efficiency which is achieved in practice. When all of the relevant factors are taken into account (such as percentage of total solar radiation received by the plant, light which is reflected or transmitted by the leaves and light absorbed but converted into heat), the final efficiency of energy conversion into usable carbohydrate is something like 0.2 percent. Even though this may seem to be a very small figure it still means that an average of 87×10^9 tons of sugar are produced by photosynthesis

per year! (The annual mining and manufacturing product of the earth is about 1×10^9 tons.)

The utilization of carbohydrate

Plants use the carbohydrates that they produce by photosynthesis in order to grow. Some of the carbohydrates are used directly in order to make cell walls and the main structural components of the plant, some are kept for future use in special storage organs or as components of seeds as food reserves and others are utilized directly to provide energy for the multitude of synthetic processes which the plant employs in vegetative growth. The plant releases energy stored in carbohydrates through the process of respiration. This process is essentially the opposite of photosynthesis.

The energy released is 'trapped' in the form of 'high-energy substances' which the plant then uses to manufacture essential compounds needed for growth. Just as plant cells contain special structures concerned with photosynthesis, they also contain special structures (mitochondria) concerned with respiration.

Basically, the process of respiration involves the cleavage of carbohydrates such as glucose into smaller units, which are then transported

Above left: When plants trap light energy they take up carbon dioxide and release oxygen. Bubbles of oxygen gas can be seen forming on the leaves of this water plant.

*Right: A magnified section of a maize (**Zea mays**) root. The area of cell production is limited to a region just behind the tip. The root cap forms a protective layer at the root tip.*

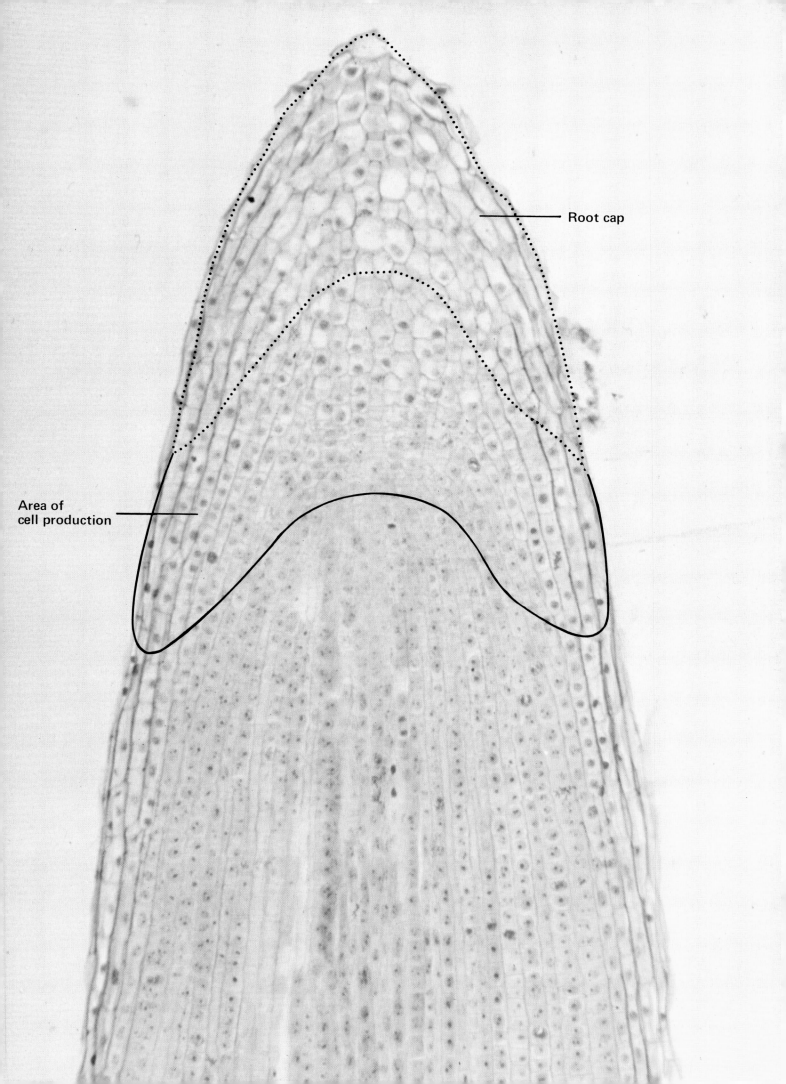

Root cap

Area of
cell production

known as DNA (deoxyribonucleic acid) contained in the chromosomes of the cell nucleus, followed by the division of this material and its containment within two new daughter cells formed when the original cell divides into half. Continued growth of the two new daughter cells then allows a repeat performance of the whole sequence.

The importance of water and its movement in the plant

Between about 80 and 95 percent of the weight of plants is water. The continued growth of a plant is totally dependent on water which in a typical herbaceous species is obtained from the soil. Of the water taken up by the roots about 98 percent is lost again by diffusion through specialized pores in the leaves (called stomata) and to a much lesser extent through the leaf outer skin or cuticle.

The diffusion of water through stomata in the leaves occurs by a process called transpiration. Stomata, although variable in size, are very small (from about 5 microns × 3 microns to 40 microns × 10 microns) but are still large in comparison to the size of the water molecules that pass through them. Leaves contain large numbers of pores on both their upper and lower surfaces but there are usually more on the lower than the upper leaf surface.

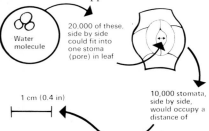

Stomatal openings in leaves can open or close depending upon a number of environmental factors. A closed pore prevents the escape of any water from the plant but will also stop the entry of carbon dioxide into the leaf thus stopping photosynthesis. A wide open pore, while allowing ingress of carbon dioxide and thus permitting photosynthesis, will also permit an easy loss of water from the plant which, unless replaced by further uptake from the roots, will cause the plant to wilt and droop. In practice progressive stomatal closure tends to reduce water loss more than photosynthesis and so many plants partially close their stomata enabling them to minimize water loss while having little effect on photosynthesis.

The main factors that change transpiration rates are temperature, atmospheric humidity, availability of soil water and light. Light has an indirect effect since it can alter both the temperature and the extent of stomatal opening. The other factors influence transpiration through their effects on changes in the water vapour pressure gradient between the leaf and the atmosphere. The steeper the gradient, the greater the transpiration rate, and vice versa. Thus on a warm dry day with a slight breeze transpiration rates would be high, provided that there was no shortage of soil water available to the plant.

Water uptake using cohesion forces
Any loss of water through transpiration has to be made up by a simultaneous uptake of water through the roots. Most plants have extensive and deep root systems which enable them to extract the water from a large volume of soil.

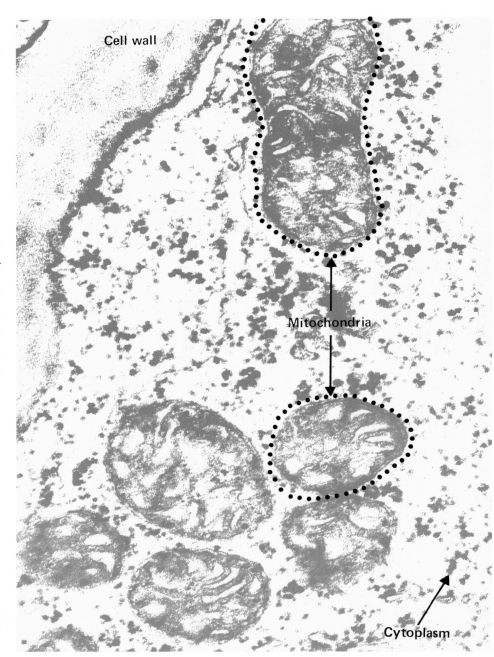

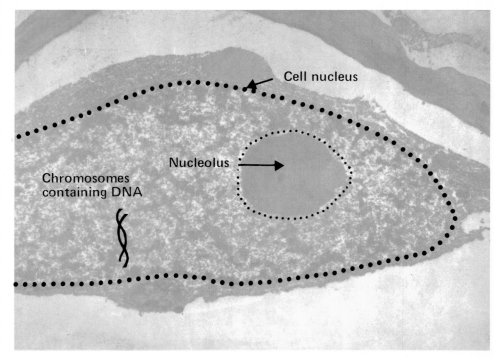

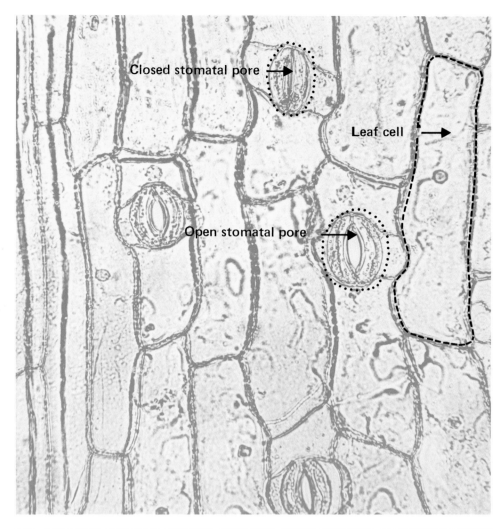

Closed stomatal pore

Leaf cell

Open stomatal pore

Not only are plants able to take up this water but they can raise it to the upper parts of the plant—a height of perhaps 100m (328ft) in the tallest trees such as the redwoods of the Californian forests. How is it that water can be carried to such heights which are many times that which the atmosphere will support through its own pressure (about 1,000cm—394in)?

There is no completely satisfactory explanation but the best idea seems to be that of the so-called cohesion theory. This idea is based upon the small tubes (called xylem elements) which make up part of the plant structure and provide a continuous 'pipeline' from the root tips to the top of the plant. When water is lost through transpiration at the top of the plant more is drawn up through the xylem from the roots which in turn take up more from the soil. The plant acts like a wick with the water columns being held together by cohesion (the attraction of like molecules for each other).

The existence of root pressure

There is evidence that under certain conditions instead of water being drawn up the xylem it can be pushed up by the roots acting as a pump. If a plant is healthy and well-nourished its roots can accumulate large quantities of mineral salts. The presence of these salts in the plant cells increases the ability of these cells to take up water which is then forced up the xylem elements as part of the transpiration stream. Even though such root pressures can be experimentally observed recorded pressures are not high enough to account for water movement to the top of the tallest plants.

The mineral requirements of a plant

Xylem elements not only carry water but also various minerals which the plant obtains from the soil. A growing plant requires most or all of sixteen chemical elements; nine in large amounts and seven in small. Of the former (called macronutrients), carbon, hydrogen and oxygen are derived from carbon dioxide and water. The others, nitrogen, phosphorus, potassium, sulphur, calcium and magnesium are obtained from the soil (except in a few species, such as clover, which obtain their nitrogen from the air).

The other seven nutrients which are required for plant growth are termed micronutrients or trace elements since they are only required in relatively small quantities. These are iron,

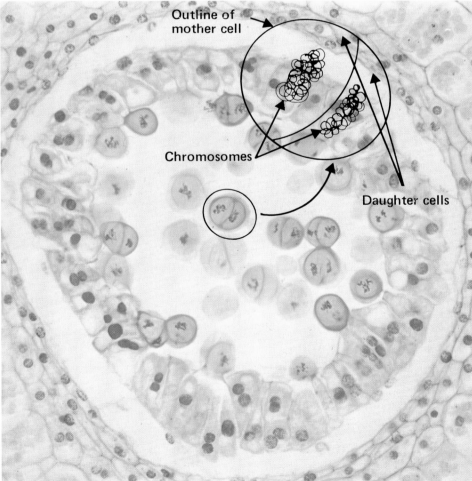

Outline of mother cell

Chromosomes

Daughter cells

Far left, above: Plants obtain the energy needed for growth by respiration of sugars. The respiration process takes place in special structures called mitochondria. This picture shows the appearance of mitochondria in Jerusalem artichoke (Helianthus tuberosus) cells.

Far left, below: A picture of the hereditary material (DNA) which is contained in the nucleus of a Jerusalem artichoke (Helianthus tuberosus) cell. The nucleolus has a special function but also contains DNA.

Above left: Stomata in leaf cells. The sausage shaped 'guard' cells can change their shape to open or close the pores according to air moisture levels.

Left: Cell divisions occurring during the manufacture of pollen grains in a lily (Lilium sp.).

manganese, boron, zinc, molybdenum, chlorine and copper.

Plants take up mineral nutrients through their roots from a solution in soil water. The required nutrients must either be in, or capable of being transferred into, a soluble form ready for transport into the plant otherwise the plant cannot use them. For example, it is quite possible for cereal crops to starve in soils which are amply supplied with phosphorus and potassium simply because the latter are in an insoluble form unavailable to the plant.

Although mineral nutrients might be used in many ways to sustain the growth of a plant, the absence of any one required mineral nutrient results in the production of certain characteristic growth deficiency symptoms. These include a reduction of chlorophyll synthesis giving the plant a pale green or yellow colour (chlorosis); localized death of tissues such as buds, leaf tips or edges (necrosis); production of red colours in stems or other structures where it is usually absent; stunted growth; unusually thin and woody stems; poor reproductive development; and sometimes the complete failure of fruit or seed development.

Phloem: the plant's arteries

Not only do water and minerals have to be moved upwards through the xylem pipeline but also material made in the leaves by photosynthesis has to be transported about the plant to the various places where it is required for growth. Plants contain very special cells whose job it is to perform this function. These cells are called sieve-tubes and occur in the phloem tissues. Sieve-tubes are only found in angiosperms. Other cells of a very similar nature perform the same job in the gymnosperms and lower vascular plants.

Sieve-tube elements contain no nucleus, but instead they are associated with other cells called companion cells which lie parallel to the sieve-tubes and which are interconnected in such a way that the loss of the nucleus is overcome. The ends of sieve-tube cells are perforated to form sieve plates and are connected up longitudinally so that a 'pipeline' is formed through which organic material can be carried. The transported material is mainly sucrose which occurs in the phloem as a 10 to 20 percent solution. Other materials are also transported, but in smaller quantities. Thus amino-acids, minerals, plant growth hormones and other molecules may all be found in the translocation stream. There is, however, a mystery concerning the way that sieve-tubes work.

The most accepted idea, although not completely satisfactory is known as the Munch 'mass' flow hypothesis after the botanist who proposed it. He suggested that the 'producer' cells in leaves which contain a high concentration of material are caused to take up water from the surrounding tissues. This causes a flow of material through the phloem tissues to the 'receptor' cells where it is then used. The excess water passes into the xylem tissues where it rejoins the transpiration stream. It is a kind of circulatory system.

The integration of structure and function in leaves

The leaf as the main site of production of material for plant growth is obviously a very important part of a plant. The main function of leaves appears to be that of photosynthesis and their structure is well adapted for this purpose. Leaves are usually thin and flat presenting large surface areas for maximum ability to absorb light energy. In certain cases the shape may be

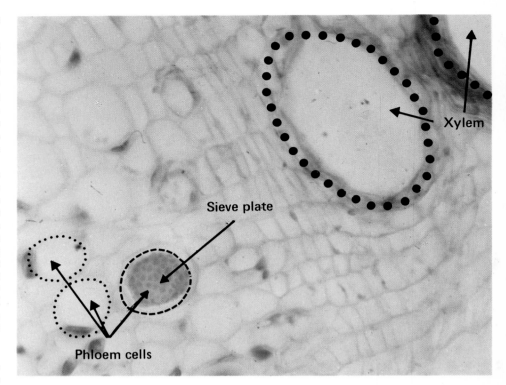

Xylem

Sieve plate

Phloem cells

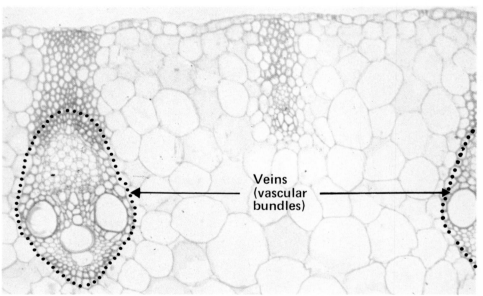

Veins (vascular bundles)

modified for specific environmental reasons. For example, succulent plants from desert areas very often have thick fleshy leaves with a thick waxy cuticle designed for maximum water storage with minimum water loss; many conifers have needle-shaped leaves designed to minimize water loss during the cold season when the ground is frozen and also to minimize snow accumulation thus reducing the snow burden on the branches.

The internal structure of the leaf is also designed to allow for maximum photosynthetic efficiency. There are large spaces within the leaves adjacent to stomata which give a high surface area for absorption of the carbon dioxide required for photosynthesis. The large surface area also allows for the easy removal of oxygen produced as a by-product of photosynthesis.

The products of photosynthesis are carried away to other parts of the plant where they are needed for growth by a very extensive vascular (or transport) system. This system is made up of large numbers of veins which permeate the entire leaf blade.

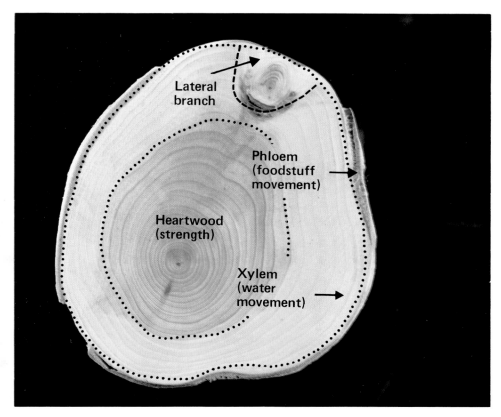

Lateral branch

Phloem (foodstuff movement)

Heartwood (strength)

Xylem (water movement)

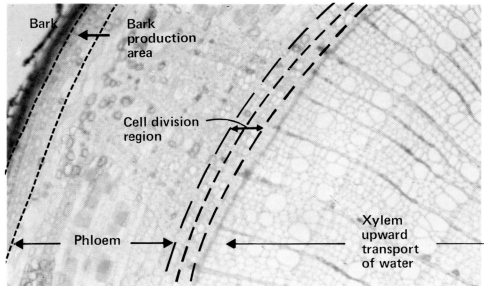

Bark

Bark production area

Cell division region

Phloem

Xylem upward transport of water

Far left, above: The appearance of phloem and xylem cells in a section of a cucumber plant (Cucumis sativus). The end wall of a phloem cell called the sieve plate can be seen.

Far left, below: A section showing the appearance of veins in a cereal leaf (maize) (Zea mays). The veins are actually arranged in a parallel fashion along the length of the leaf.

Left: A section of wych elm (Ulmus) stem. The annual rings representing different rates of growth throughout the year can be seen very clearly.

Below left: A close up of the cell division area in a lilac (Syringa sp.) stem. The two areas of cell division produce three kinds of cells, xylem, phloem and bark (cork).

In most plants elongation growth takes place in regions just behind the tips of shoots and roots. In shoots this region of elongation growth extends some distance down the shoot but in roots is completely restricted to the area immediately behind the tip.

Growth in different directions—increase in girth

Plants not only grow upwards, but outwards. Outward growth occurs by the production of lateral branches (giving a plant its characteristic shape) and also by the thickening of the stem. The mechanism whereby this latter growth takes place is more easily understood if we consider in a little more detail the structure of the stem in a woody plant which contains secondary meristems.

A stem is made up of several concentric layers of cells surrounding a central region of older tissue. All of these different layers are composed of cells which have different functions. The innermost layer is composed of xylem tissues through which water is translocated to the leaves. Adjacent to this layer is a meristem known as the vascular cambium. This is a region of dividing cells some of which give rise to new xylem cells and some of which produce cells for the next outer layer which is composed of phloem tissue. Adjacent to this there is sometimes another meristem known as the cork cambium and from this the outermost protective 'skin' or bark of the plant is produced.

Increase in girth is quite simply the result of successive cell division in the vascular cambium giving rise to new xylem and phloem tissues. As this increase in lateral size takes place, the layers of cells split apart as they are pushed outwards. The vertical gaps so produced are filled in by the activities of the cork cambium and cambial tissue within the stem.

Growth rings in woody plants

In temperate regions there is a marked seasonal periodicity of growth, both in the primary growing points at the apices and in the secondary growth from lateral meristems. Growth is usually very rapid once initiated in the spring but at the apex slows down quite quickly where it has usually stopped by the summer. Secondary growth or increase in girth also slows down but continues until the autumn when the plants finally become dormant.

It is this periodicity of growth that causes the production of growth rings, the study of which can provide considerable information on the climatic conditions and local environment

The veins not only remove the products of photosynthesis from the leaf but also furnish the leaf with necessary water, essential minerals and certain organic materials manufactured in the roots and older leaves. It is an extensive two-way transport system ensuring the efficient and rapid utilization of photosynthetic ability for the maintenance and further growth of the plant.

Growth and the formation of meristems

Growth implies an increase in size and volume. It can also be taken to mean an increase in complexity and then it is perhaps more properly described as differentiation. In plants these two aspects of growth are seen quite separately but both are ultimately dependent upon the activity of special regions called meristems. These are areas of embryonic tissue from which all of the

Left: Netted venation in a leaf of black poplar (Populus nigra). The veins have been exposed by natural decay processes in a stagnant pond.

new tissues of a plant are generated. All plants have meristems which are located at the tips or apices of the shoots and roots. For many plants, particularly herbaceous forms (ie plants having little or no woody tissue), they are the only meristems present and because of their position these are known as apical meristems.

In woody plants other meristems are present also, known as secondary meristems. The cells produced in these regions are responsible for increase in girth, for example the increase in width which takes place when a tree grows.

Growth in different directions—elongation growth

Elongation growth in plants is achieved through cell division followed by cell enlargement. Growth rates vary enormously; for a plant growing at its maximum rate an increase in height of about 1–5cm (0.4–2.0in) per day is normal but values as high as 60cm (24in) a day have been recorded, eg in some species of bamboo.

Growth is not uniform over the whole plant.

experienced by a tree. Usually only one ring is formed per year but under certain conditions such as insect infestation or a late frost causing premature defoliation and a subsequent interruption of growth, two rings per year may occur. This latter situation can usually be detected since the rings so produced are likely to be unusually narrow.

Support by woody tissue

The increase in girth provided by the activities of the lateral meristems is the result of a thin cylindrical layer of cells continually being created in the outermost layers of the stem. What happens to the older cells?

The older cells of the phloem eventually die and can no longer be used for transporting organic solutes. They become incorporated into the outermost corky or bark layers of the tree. The older cells of the xylem which form the sapwood of the tree eventually become non-functional and generally become impregnated with a variety of organic substances such as tannins, gums, oils and resins. These cells then form the heartwood of the tree. The primary function of the heartwood is as a mechanical support and the changes which occur in its formation contribute towards increased compressional strength and confer an increased resistance to decay.

The control of growth—plant hormones

Plants are very sensitive to changes in their environmental conditions and a change in external surroundings can have a profound influence on the rate or type of growth. Despite this, however, growth does not take place in a haphazard way but is controlled very closely by the plant itself. Growth does not occur uniformly all over a plant but in certain specific regions. Growth between specific regions can be correlated too—for example, a root never completely outgrows a shoot or vice versa. One obvious explanation for this is a nutritional one. Roots and shoots are interdependent because shoots need the materials extracted from the soil by roots while roots need the photosynthetic products of shoots for their continued growth.

The situation is more complex than this and it is now known that rates of growth in plants are controlled by special messenger molecules or hormones. These compounds are generated at various places within the plant and transported in very small quantities to their sites of action where they can profoundly influence the type of growth taking place.

The discovery of plant growth hormones initiated entirely new areas of experimentation which are still being pursued and although hormones are now produced and used in quantity for various commercial purposes, such as seed set, fruit ripening and breaking of dormancy, their modes of action are still imperfectly understood.

Day length and other influences

Perhaps the two main influences the environment has upon plants are those on flowering and movement. Quite often these effects are mediated through the action of growth hormones.

In angiosperms, flowering is very often controlled by the length of the daily period of illumination. This response is known as photoperiodism and normally day length has to fall within certain limits for flowering to occur. Three main groups of plants can be identified. Plants which flower if the day length is less than a certain critical value are called short-day plants, eg chrysanthemums, xanthium and strawberries. Plants which flower if the day length exceeds a certain critical value are called long-day plants, eg henbane, spinach, lettuce, wheat and beet. The third group of plants are called day-neutral plants and these can flower at any time regardless of the day length, eg tomatoes, dandelions and sunflowers.

The way in which flowering is controlled by daylight is not fully understood. It seems however that plants contain a pigment called phytochrome which changes its nature in the presence of light.

This pigment is able to control the movement of a hormone or group of hormones from the leaves to the apex of the plant where the flowering process is initiated.

Movements in plants are brought about either by the unequal distribution of growth or by changes in cell water content. The direction of movement may be related to that of the application of the stimulus, in which case it is called a tropism, or it may be dependent on the plant alone, known as a nastic movement.

The two main tropic movements which can be observed are those of phototropism and geotropism. Phototropism is the bending of a shoot towards a light source. The magnitude of the response depends upon the amount of light received and thus the same effect can be produced by low intensity light given for a long period of time or by high intensity light given in a flash. The effect is caused by the unequal distribution of a hormone, called auxin, following the application of light. The auxin concentrates on the shady side of the plant where it causes greater cell growth and hence a bending of the plant towards the light.

Auxin is also involved in geotropism. Geotropism is the response of plants to the earth's gravitational field, which causes shoots to grow upwards and roots to grow downwards. Since they grow directly towards the stimulus (the earth) roots are said to be positively geotropic whereas shoots, growing in the opposite direction are negatively geotropic.

The other sorts of movement in which the direction of movement is independent of the direction from which the stimulus is received are the so called nastic movements or nasties. The principal nastic movements are those caused by changes in light intensity (photonasties), temperature changes (thermonasties), touch (thigmonasties or seismonasties) and alternation of day and night (nyctinasties).

The rapid opening of a flower when it is brought from a cold room to a warm one is an example of a thermonastic response, while the opening of tobacco flowers in the evening is a photonasty resulting from a decrease in illumination intensity. Some of the most common examples of nyctinasties are the 'sleep movement' of plants such as the raising or lowering of the leaves of a bean plant during the day and night to give a horizontal or vertical orientation respectively.

Perhaps the most dramatic responses, however, are the thigmonasties. When an insect settles on the leaf of the Venus' fly trap (*Dionaea muscipula*), the blades fold together, the spines on its margin interlocking to imprison the hapless victim. Another example is the sensitive plant *Mimosa pudica*. When this plant is stimulated water escapes from special cells at the base of the leaf stalks causing the leaflets to fold up in pairs and the main leaf stalk to drop. If a very strong stimulus is applied to one leaflet it can be transmitted throughout the whole plant at the rate of about 3cm (1.2in) per second.

Plant shape: making the best use of space

All plants have characteristic shapes which have been determined throughout the course of evolution to best fit them for survival in the particular environmental niche which they occupy. There are many environmental factors which contribute towards the shape that a plant takes; eg desert cacti have evolved their peculiar and characteristic shapes in response to the need to conserve water and yet be able to photosynthesize. Thus their leaves have been reduced in the course of time to no more than needles or spines, while the main body of the plant, where all the photosynthesis now takes place, has become swollen for purposes of water storage.

In a mixed deciduous forest the shapes of the trees are considerably modified by the pressure of space. Close packed trees tend to be tall with few lateral branches in the effort to obtain as much of the sunlight as is available by outgrowing all their neighbours, whereas well spaced out trees assume a more symmetrical and well proportioned shape being able to spread their lateral branches out to a greater extent.

The problem of obtaining sufficient light for photosynthesis is a very important factor in determining plant shape and leaf arrangements play important roles in determining the photosynthetic efficiency. For example, the mode of growth of a cereal crop enables a good photosynthetic efficiency to be obtained. In cereals the leaves grow from the base of the plant assuming a nearly vertical angle of growth. In this way all of the leaves receive some light and are able to make a direct contribution to photosynthetic efficiency, although the effective light intensity is reduced because the leaves are not orientated at right angles to the sunlight.

Many plants which have horizontally placed leaves have evolved a shape which places the leaves in a spiral pattern. This gives the maximum chance of intercepting light with a minimum of shading. In addition, other plants are able to either move their stems or their leaves so that a maximum leaf area is presented to the incident light.

Light is not the only environmental factor which influences shape. Temperature, wind, soil conditions, salt spray and even grazing animals all play their part. For example, an oak tree growing in a sheltered forest appears very different to the stunted and tortured structures which are assumed on windswept and colder moorland areas, and many trees growing in coastal areas are bent and forced to grow in a direction away from the prevailing winds.

However, whatever variety of external shape and form is forced upon a growing plant by the environmental conditions it experiences, it is always recognizable as a particular species because the fundamental aspects of its structure are still determined by its inherited genes in a way not properly understood and which forms one of the great mysteries of plant science today.

Right: Young and old leaves on a cinnamon tree (Cinnamomum sp.). The areas of new growth can be distinguished by the reddish colour on both leaves and stem.

SOME IMPORTANT PLANT GROUPS

Helleborus niger
Christmas Rose

SEAWEEDS AND ALGAE
first plants of sea and land

What exactly is an alga? A simple answer to this question has eluded botanists for centuries and even with the help of modern research the answer remains highly complex. Remember that the plant kingdom has been classified by grouping together species which show common features. How then can one account for a single group containing, on the one hand, microscopic plants composed of a single cell and swimming freely in water, and on the other hand giant seaweeds over 100m (328ft) in length, which may lay claim to being the longest-living things on earth? The position is further complicated by some zoologists who claim the single-celled species, which swim by means of whip-like flagella, as animals!

Perhaps the simplest way to think of the group is not so much what they have in common, but rather what features, as a group, they do not have when compared with higher plants, as what does typify the group is their relative simplicity.

Algae have never evolved the complex root and water plumbing system of higher plants and so have been unable to colonize dry land to any extent. They are plants of fresh or marine waters with seaweeds as the most obvious representatives in the intertidal and shallow water zone around every coastline. Here they are also able to survive without water-regulating features such as cuticle and stomata. Some of the most simple algae are able to colonize soil, rocks and tree trunks, but are only able to flourish so long as there is a film of surface moisture. When conditions become too dry, they quickly become dormant until suitable conditions return.

Although the photosynthetic pigment chlorophyll is found in all algal groups its characteristic green is often masked by others. These give some species a distinctive colour which was used in early attempts at classification. In particular, it was used to divide two of the major seaweed groups—the red and brown. The key to a modern classification of the algae has been the development of techniques for studying their biochemistry and cell structures. These studies have upheld the validity of dividing red and brown algae. The chemical responsible for the red colouring is the biloprotein, phycoerythrin, while the browns are coloured by a carotene, fucoxanthin. The red and brown algae (Rhodophyta and Phaeophyta) now form two of the ten divisions which are generally recognized. A third major division is the green algae (Chlorophyta) where chlorophyll remains dominant.

One division, the blue–green (Cyanophyta), is composed of species which have cells quite unlike those of other algae. These cells, known as 'procaryotic' cells, are much more like those of bacteria and appear much simpler in organization. One of their unique features is that no true nucleus is present as in the normal 'eucaryotic' cells of other algae and all other plants. Because of their simplicity the procaryotic cells are thought of as being more primitive than eucaryotic ones. There is fossil evidence for this, as algae found in Precambrian rocks are thought to be of the blue–green type. These plants must have been alive over 600 million years ago when life on earth was still at a very simple stage. Speculating about the evolution of any group of living things is a fascinating exercise and the algae are no exception.

Form and reproduction

Chlamydomonas—a unicellular alga

A unicellular organism is one which consists of a single cell, in which occur all the life processes—respiration, reproduction and, in the case of most algae, photosynthesis. The 325 or so species of Chlamydomonas occur widely in freshwater and are especially associated with water rich in ammonium compounds. They also occur on damp soil. The cell structure of Chlamydomonas has many features in common with cells found in higher plants, but the special features which allow Chlamydomonas to live as free-

Fucus vesiculosus
Bladder Wrack

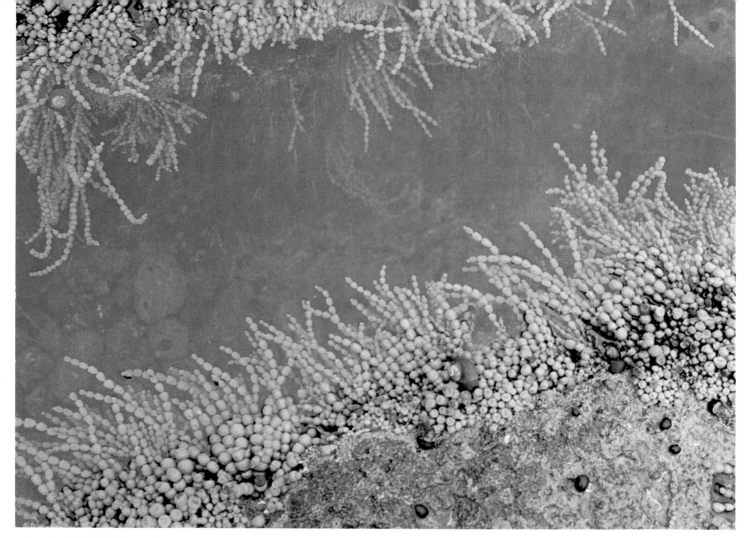

swimming individuals are the two, or sometimes four, flagella, a light-sensitive red pigment or 'eye spot' and a single cup-shaped chloroplast. Reproduction can be either sexual or asexual. Asexual reproduction is much the more frequent phenomenon, allowing *Chlamydomonas* to increase in numbers very rapidly. In this case the protoplast of an individual may divide several times within the parent cell wall to produce up to 32 daughter cells. Sexual reproduction occurs when two individuals act as sexual units or gametes. These then join to form a single unit known as a zygote. When in the zygote resting stage a thick wall protects the contents from drought; in this condition the zygote may be blown from place to place. When one lands in a suitable place development is completed and four new individuals are released and become mobile once more. This resistant zygote stage which results after sexual reproduction is a common phenomenon throughout the algae except for the blue–green, where sexual reproduction has never been observed. Although in most species of *Chlamydomonas* the two parents of the daughter cells are apparently identical, in other species distinct male and female parents are discernible.

*Above: The Venus' necklace, **Hormosira banksii**, is an exposed shore seaweed restricted to Australasia. Like all seaweeds it has a complex structure, in contrast to green terrestrial algae.*

*Right: The terrestrial alga **Pleurococcus** forms a green film on trees and stones. It is a single-celled alga, one of the few species able to exist away from water.*

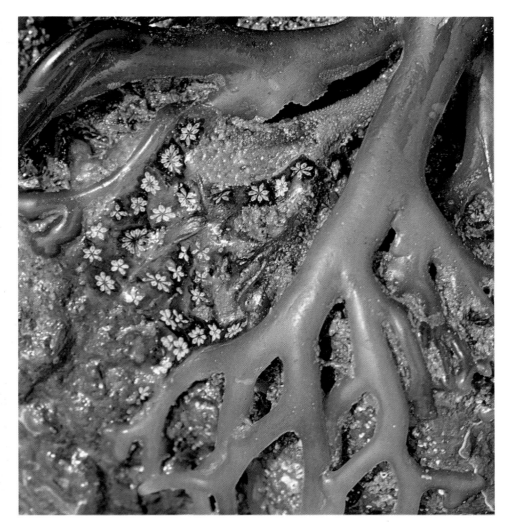

Above: Oarweeds like **Laminaria digitata** anchor themselves to rocks with powerful holdfasts, which are their substitute for roots. The little stars belong to a sea squirt.

Left: Only at extreme low tides are the tough, rubbery fronds of the oarweed **Laminaria digitata** exposed to the air. European oarweeds reach about 5m (16ft) long, but some kinds exceed 200m (660ft).

Volvox—a colonial alga

Colonial algae usually rather resemble a mass of unicellular algae embedded in a gelatinous mass. In fact, the individual cells of *Volvox* look remarkably like *Chlamydomonas*. The disorganized appearance of the colony is misleading as its number of cells is determined when it is formed and does not increase during the life of the colony. A colony may contain as many as 20,000 cells each bearing two flagella. These cells are contained in a gelatinous sheath in the form of a sphere. The spheres which occur in freshwater ponds are just large enough to be seen by the naked eye and may be numerous enough to colour the water green. Movement of the many flagella is coordinated, so that the whole colony is able to move slowly through the water.

Asexual reproduction occurs by the formation of a number of daughter colonies within the sphere, which are then expelled. Sexual reproduction is accomplished by the formation of free-swimming male gametes known as antherozoids. These fuse with egg cells which form individually within the sheath. As in

Chlamydomonas, a resting zygote is the result of the union.

Spirogyra—a filamentous alga

This well-known alga is widely distributed in freshwater ponds where its unbranched strands may form a dense mat sometimes referred to as blanket weed. Each filament consists of a row of cylindrical cells placed end to end. Within the protoplast of each cell is situated one or more of the distinctive spiral chloroplasts which give these algae their name. Each strand grows longer by cell divisions. These are in one plane only so that the filament remains a single row of cells. Accidental breakage of these filaments appears to enable new filaments to grow and provides a means of asexual reproduction.

Sexual reproduction occurs when two filaments come to lie side by side in a gelatinous mass. Cells from each of the two filaments produce outgrowths of the cell wall which link with the cell in front. At this stage the two filaments look remarkably like a minute ladder. These connections are known as conjugation tubes. No flagellate gametes are produced, as the contents of one cell merely flow into the other where the zygote is formed.

Ulva—a parenchymatous or leafy alga

These green seaweeds all occur in the marine mid-tidal zone where they are known as sea lettuce. Unlike the algae already described, *Ulva* contains cells of different kinds, specialized to form a 'leaf' and holdfast. This flat 'leaf' or thallus is produced by division of cells in two or three planes, while the holdfast is composed of colourless rhizoids which arise from certain cells at the base of the thallus. On emerging from the thallus they become matted together to form the holdfast which anchors the plant

to the rocks. Although the holdfast acts as a true root in functioning as an anchor, it does not function as a water- or nutrient-absorbing organ. A new holdfast can be produced from a fragment of thallus, but this form of asexual reproduction does not appear to be common.

Sexual reproduction, which is the norm in *Ulva*, provides us with an example of alternation of generations which is not only a feature of the algae, especially seaweeds, but also in modified form of all higher plants. In *Ulva* both of these generations produce free-swimming single-celled 'swarmers'. These generations are the haploid gametophyte producing true gametes, two of which fuse to form a zygote, and the diploid sporophyte producing zoospores by meiosis. These do not fuse with one another but develop directly into a new haploid plant after a short resting phase. Swarmers of both types may be liberated in such vast numbers as to colour the water green. Along the Californian coast the release of swarmers occurs bi-weekly giving regular 'green tides'.

The stoneworts (Charophyceae)—specialized parenchymatous algae

Members of this specialized class of algae may easily be confused with higher plants as their thalli, usually about 20 centimetres in height, consist of distinct nodes and internodes. They occur in fresh and brackish water, although they thrive best in lime-rich water. Here they may completely encrust themselves with lime or 'marl'. Not only is the thallus complex in structure, but the organs producing the sexual cells are also highly specialized. Although mobile male gametes are produced, the egg cell remains within the parent plant until after fertilization.

Red and brown algae (Rhodophyta and Phaeophyta)

All the preceding examples are members of the green algae (Chlorophyta), which have been used to illustrate some of the range of structures and methods of reproduction found in the algae. Such a range of structures can be found in other divisions, but it is in the red and brown algae that the most specialized parenchymatous forms have evolved. It is these which form the majority of true seaweeds. Although the red algae have evolved the most complex reproductive systems, it is the brown algae that have developed the most elaborate plant forms. The group includes such well-known seaweeds as the giant kelps *Nereocystis,* the various rockweeds or wracks of the genus *Fucus* and the floating seaweeds *Sargassum*. Both divisions exhibit forms of alternation of generations.

Ecology and economic uses

Seaweeds in general are very specific in their environmental requirements. The brown algae, for instance, are typically seaweeds of cool waters including the Arctic and Antarctic, while the red are more abundant in warm waters. Each species is also usually associated with a particular tidal band or depth of water. The brown alga, *Fucus,* prefers the higher tidal belts, while *Nereocystis* is characteristic of deeper water. Both species are anchored by means of a holdfast. As some species of giant kelp may be over 100m (328ft) in length, the holdfast is often a substantial structure. *Sargassum filipendula,* on the other hand, is freefloating and collects in vast floating mats such as in the notorious Sargasso Sea in the mid-Atlantic.

Although of such lowly status in the plant

kingdom algae may significantly affect the environments in which they live. The lowly unicellular and colonial algae form the plant components of plankton and as such form the vital first stage in the food chain of marine animals. This vast mass of simple plant and animal life found floating near the surface occurs not only in the sea but also in freshwater lakes. A number of species of whale live entirely by filtering plankton from the oceans. Some of the larger brown and red algae are more directly of benefit as a human resource. Some of the chemicals extracted from seaweeds include colloidal gels like pectin, carragheen and agar, as well as iodine from various red algae. One of the most important gelling agents extracted from seaweed is algin, which has a wide range of industrial uses: its salts are used to 'dress' textiles, thicken dyes, join tweeds in weaving, glazing paper, as a suspending agent in the preparation of medicines, paints, cosmetics, insecticides, etc, for waterproofing fabric, and as a stabilizer in ice-cream. It is harvested commercially from species of *Laminaria* along a number of European coastlines. It is off the Pacific coast of the USA, however, that the industry has become most mechanized. Machines, rather like giant combine harvesters, cut and scoop 300tonnes of the giant kelp *Macrocystis* per day.

Because of its economic value there have been a number of proposals to introduce the productive American giant kelp *Macrocystis pyrifera* into European waters. This has caused alarm among marine biologists who consider that this productive seaweed could cause profound changes

in the eco-system of the shallow waters around the coast, replacing the native *Laminaria* and affecting fish and shellfish grounds. A lesson has already been learnt from the accidental introduction of the Japanese seaweed *Sargassum muticum* to the USA and elsewhere. This has caused considerable problems and its recent appearance on the south coast of the United Kingdom has been regarded with alarm, and attempts are being made to eradicate it.

Algae can be less than beneficial in other ways. Those which form the plankton have an enormous capacity for growth and where conditions are right, 'blooms' occur which may colour water the colour of the dominant algae. Organisms present in such vast numbers may cause many problems. Their waste products may cause chronic poisoning of fish and other marine life. Today many freshwater lakes are undergoing eutrophication—enrichment by sewerage effluent or nitrogen fertilizers which have been washed off adjacent farm or forestry land. In such nutrient-rich conditions algae flourish. Although during the day the oxygen produced by photosynthesis masks that used for respiration, at night oxygen in the water is quickly used up by the algae, often causing extensive fish deaths through suffocation.

Evolution

The relative simplicity of algal structures makes them an obvious group in which to look for the ancestral forms of higher plants. It is easy to imagine the range from simple to relatively

Above: The extraordinary sea palm, **Postelsia palmaeformis**, *off the Californian coast. The stout stem allows the oxygen-carrying water to flow freely around the long flexible fronds.*

Right: Planktonic algae like **Euglena** *may form 'blooms' when the conditions for maximum increase are right. The coloured scums that result may poison other marine life.*

complex structures found in a number of divisions as stages in an evolutionary sequence leading through to higher plants. There is certainly evidence for this in the case of Chlorophyta which like higher plants contain chlorophyll as the dominant photosynthetic pigment. As the first fossil vascular plants occur in rocks of Silurian age over 405 million years ago, this split must have occurred even earlier and it is thus most unlikely that forms showing direct ancestry would have survived. Similarly, the algal divisions themselves may have had a common ancestor but each division has evolved in its own way. Evolution in the algae has produced a number of complex forms, some of which show parallels with the evolution of higher plants. These include cell differentiation to perform specific functions within the plant, including translocation. Sexual reproduction has been refined to a point where the female cell is retained in the parent plant until after fertilization, and alternation of generations, fundamental to the evolution of higher plants, is present.

Cradle of the deep~the sea otter and kelp

The Californian sea otter lives permanently in the great beds of kelp off the coast—it even sleeps there, wrapping the fronds round its body for anchorage, sure of reasonable comfort because the kelp damps down wave motion to a considerable degree. Many invertebrates also live among the seaweeds, notably small molluscs and crustacea which the otter eats. The sea otter does in fact maintain a natural balance, because these small animals feed on the kelp, and where the otter has been destroyed by man the kelp gradually disappears as a result.

Between 1910 and 1930 kelp was harvested off the western coast of U.S.A. as a rich natural source of potash, but this finally proved uneconomic, as did wartime production of acetone. The major value of kelp today, in California as elsewhere, is in producing alginates, which have many industrial applications.

MOSSES
simplest of land plants

Mosses are the simplest true land plants which are able to withstand drought and grow in dry habitats, such as rock faces and stone walls. However the vast majority of species are to be found associated with freshwater along stream sides, in bogs or in moist woodland and tropical rainforest. The 14,500 species of moss are typically small, low-growing or mat-forming plants a few centimetres in height. The Australian genus *Dawsonia* may, however, reach 70cm (27.5in) in length. In northern Europe, one of the largest mosses *Polytrichum commune* may grow to 20cm (9in) in acid bogs, while in New Zealand a length of 150cm (59in) has been claimed for the same species growing under water. A number of species such as *Fontinalis antipyretica* are only found in running or still water and this species may grow to almost 100cm (40in).

Structure and life cycle

Before we go further in studying the typical moss plant and comparing it with the structures seen in higher plants we must be aware of one important difference. This is a result of the phenomenon of alternation of generations, first seen in the algae. In mosses the generation which forms the typical moss plant is the haploid gametophyte. The spore-producing diploid sporophyte is normally comparatively small, grows from the leaves of the gametophyte and is partly parasitic on it. In ferns and flowering plants the obvious plant body represents the sporophyte generation, while the gametophyte has become reduced. This important difference should always be remembered when comparing mosses with higher plants.

Taking the spore as the starting point of the life cycle, this first germinates into a delicate branched filament rather like a filamentous alga. This is known as the protonema. When this has developed sufficiently, buds are produced from which the plant proper grows. This consists of a distinct stem and leaves. The stem, which may be branched, has small root-like multicellular rhizoids at its base which have a water-absorbing function. A strand of water-conducting tissue is also present in the stem. The leaves are usually only one cell thick except for a distinctive midrib which is seen in some genera including *Bryum* and *Mnium*. In *Polytrichum* the midrib occupies much of the leaf and the single-celled blade is reduced to a narrow strip of cells on either side. A cuticle is also present in these genera. Here we see that a simple system for translocation (movement of water and food supplies) has developed, although it is of rather limited efficiency, for if the surrounding air is too dry, wilting will still take place. To compensate for this, water can also be conducted externally in leaf ducts known as capillary channels. Mineral salts are also obtained externally from airborne dust.

An important difference also exists between mosses and higher plants in their resistances to drought. While higher plants survive by their ability to retain water and a turgid cell structure, mosses are able to lose almost all their water, appearing completely dead and dehydrated. However, as soon as water is applied, they quickly reflate. This enables genera such as *Tortella*, which has no cuticle, to survive on walls which are dry for much of the year. Under experimental conditions some mosses have survived temperatures of 100°C (212°F) for short periods.

The main purpose of the gametophyte is to produce the male and female gametes for which specialized structures, the antheridia and archegonia, have evolved. These structures are also characteristic of liverworts, which with mosses make up the bryophytes. The male gametes (antherozoids) are produced in the antheridia which appear as minute sac-like structures, each borne on a short stalk and accompanied by sterile club-shaped paraphyses. The archegonia, on the other hand, are flask- or bottle-shaped, having a narrow neck of cells with a rounded base known as the venter, in which the single egg cell develops. When this is fully developed the cells in the centre of the neck break down, allowing entry of the gametes which swim from the antheridia. One of these fuses with the egg and fertilization is complete. Specialized leaves not only protect the antheridia and archegonia but also hold water to facilitate the swimming of the gametes. As, in some species, plants may be either male or female exclusively, the gametes may have some distance to swim. Water is therefore vital to the reproduction of mosses.

Once fertilized, the egg is able to develop into the sporophyte. This consists of an organ for absorbing moisture from the parent gametophyte, a stalk or seta, and a terminal capsule in which the spores are produced. As the capsule matures, the seta lengthens slowly usually to about 1–2cm (0.4–0.8in). The ripe capsule may be pear-shaped or cylindrical with a distinct cap. This cap is then shed, leaving behind a single or double ring of teeth. These are known as the peristome and function in regulating the release of spores from the capsule. Spore release in wet weather is prevented by swelling of the peristome. In some species the peristome may act by actively flicking out the spores as the peristome rings dry. The cells of the sporophyte, although obtaining water from the gametophyte by means of the absorbing body and a conducting strand, also contain chlorophyll. The sporophyte is therefore at least partly self-sufficient as it can produce its own food. Cuticle and stomata, similar in form and function to higher plants, are also present in some genera.

Sphagnum and Andreaea

Two genera do not fit into the general scheme described above. These are *Sphagnum*, the 'bog mosses', and *Andreaea*, a small reddish-black moss which forms small patches on rocks in exposed mountainous or arctic regions. Both genera show adaptations to their environment as well as other peculiarities. In both, the protonema is a disc and not a filament. From this *Andreaea* develops a peculiar rhizoid complex which anchors the plant to the rocks. *Sphagnum* on the other hand has no need of a strong foothold as it grows in a mass in bogs. No central strands are present in shoots or leaves, instead the outer layer of cells contain pores which allow the direct entry of water and the whole plant acts as a sponge. Both *Sphagnum* and *Andreaea* have distinctive capsules. In *Sphagnum* this is raised on a short stalk of the gametophyte, is round in shape and has no peristome. When ripe the walls shrink until the cap is blown off by air pressure and the spores released with explosive force. *Andreaea* has no cap at all. Instead, as the ovoid capsule dries it splits along four longitudinal lines of weakness, rather like a Chinese lantern. If the capsule becomes wet, however, the sections act as valves and the capsule is closed.

Ecology

Although mosses are often overlooked there are a few types of plant community where they form a significant element. In two examples, arctic tundra and in the various types of peat bog, mosses are dominant. The bog mosses are again of particular interest—nine or ten species each with a specific environmental requirement are involved in the various stages of development of highly acidic peat bogs. Here these plants dominate to the exclusion of most other species. This succession involves the gradual building of moss hummocks from bog pools. Specific species are associated with the wettest condition encountered in the pools while others such as *Sphagnum papillosum* are the main hummock formers. As these hummocks become drier they are colonized by heathland species such as heather, but over a considerable period these become moribund and the hummock breaks down to become a pool once more, while all around new hummocks are being formed. The growth of mosses may even raise the bog above the adjacent land surface to form a vast 'raised bog' which is convex in cross-section.

Mosses are of great use to the plant ecologist as indicator species. Certain species are characteristic of acid soil while others are only found in calcareous conditions. Others are indicators of nitrate and phosphate enrichment.

Right: The moss Pohlia nutans *has bright green spore capsules which hang downwards. It is a moss of acid soils—peat or sand, even decaying wood, but never chalk.*

Evolution

Although it is tempting to consider the bryophytes in an evolutionary sequence between the algae and ferns, there is little evidence to substantiate this as no living or fossil intermediates—'missing links'—have been found. Similarly, no intermediates exist between mosses and liverworts, and fossil evidence suggests that these groups were distinct at the end of the Palaeozoic (over 200 million years ago). There is, however, biochemical and structural evidence to link the green algae (chlorophytes), bryophytes and higher plants. An apparent link with ferns may be found in their gametophyte which superficially resembles a thalloid liverwort and bears antheridia and archegonia. An interesting modern theory suggests that bryophytes should be regarded as simplified descendants of one of the fern-like pteridophytes rather than a more advanced descendant of algae.

Right: Sphagnum mosses are often found in the wettest bog conditions. Their leaves contain large empty spaces which hold great quantities of water, and gives them the name 'drowned cats'.

Far right: Many mosses grow on the bark of trees in moist conditions: here an old lime tree by a river has its trunk and the larger branches entirely covered in dense moss growth.

The sphagnum bog

A typical acid bog in the New Forest, mainly composed of sphagnum moss. Sphagnum starts in bog pools, and after a time other species begin forming hummocks and the bog surface dries out a little. Such hummocks provide footholds for higher plants like the rushes seen here, and also the red-leaved insect-eating sundews just visible. Bogs of this sort are apt to rise steadily, sometimes above the level of the surrounding land. If eventually they dry out the sphagnum disappears. After many centuries peat is formed from such bogs.

FERNS
beautiful plants of shade and damp

A fern is a photosynthetic green plant that grows in soil or epiphytically on other plants, with a few species restricted to freshwater habitats. The basic construction is of roots, rhizomes and leaves (called fronds), interconnected by a complex vascular system comparable to that found in flowering plants. Reproduction is by means of spores, with the life cycle having two distinct phases—a sexual gametophyte and an asexual sporophyte.

Structure and life cycle

The fronds of most ferns arise from the rhizome, though in some species they grow from creeping stems. When young they are tightly curled and commonly called croziers or fiddleheads. As they mature they unfurl and enlarge, the size increase being due to both cell division and cell enlargement. The shape varies from simple strap-shaped as in the hart's-tongue fern *(Phyllitis scolopendrium)* to complex and highly divided as in bracken *(Pteridium aquilinum)*. The divisions of the more complex fronds are called pinnae.

Most ferns are small, few exceeding 2m (6.5ft) in height. There are, however, a few that grow considerably taller, up to 25m (85ft) high; these are mostly the tree-ferns. Often they have 'trunks' consisting of old frond bases with a rosette of fronds at the crown, and superficially they resemble palm trees. The few others that grow to a large size are the climbing ferns.

Some aquatic species have a diameter of less than 1cm (0.4in) and are often confused with duckweeds *(Lemna)*.

The fern plants that are commonly seen are the sporophytes, so called because they bear spores which develop in and are dispersed from spore-cases called sporangia that are located on the lower surface of some fronds. The fronds bearing sporangia are termed fertile, whereas those without are sterile. The sporangia usually occur in round or elongate clusters called sori, but in some species they are evenly distributed over part or all of the underside of the frond. When ripe, the sporangia split, and the spores contained therein are released. To the naked eye these spores appear as dust, each being no more than about 50 microns diameter, but when viewed through the microscope surface patterning is often apparent, usually spines, pores or ridges.

Damp conditions are required for the spores to germinate; they grow and form the gametophyte or prothallus which is normally short-lived. Most prothalli are thin, heart-shaped and less than 1cm (0.4in) long; a few are much-branched and thread-like. They mostly grow on rocks, soil or other plants and are photosynthetic, but some are subterranean and only grow after having formed a symbiotic relationship with a fungus. On the surface of the gametophyte, microscopic sex organs are formed, the male antheridia and the female archegonia, and it is in these that the sex cells or gametes develop. The male sperms swim to and one fertilizes the female egg to form a zygote; a film of water over the surface of the prothallus is essential for

this process. The zygote divides and grows to form a new sporophyte plant, living parasitically on the gametophyte until established.

This sporophyte/gametophyte cycle is called alternation of generations; it allows an intermixing of genetical material to take place which results in variation in the characteristics of the individuals of a species, the basis upon which natural selection and evolution operates.

Habitat

There are about 10,000 fern species found throughout the world, the majority of which thrive only in moist or wet regions not subjected to freezing temperatures: about 30 percent of the total grow in tropical rain forests,

and about 40 percent in subtropical and montane rain forests. Comparatively few grow in cool and dry regions.

Tropical rain forests are fairly dark, wet and hot, and provide a variety of habitats for ferns. Those at ground level are usually large and produce masses of spores. Their gametophytes find it difficult to become established on the rain-soaked mud, and in order to overcome this problem many of the sporophytes have developed methods of vegetative reproduction, such as bulbils or plantlets on the fronds which droop to the ground; these stand less chance of being smothered by mud than do the gametophytes. More sporophytes of these plants are produced vegetatively than sexually. A few species rooting in the ground climb up the trees

Gremmitis fasciala (left)
and G. subpinnatifide

Above: Most ferns grow in rain forests and almost all prefer deep shade, like these tree ferns (Cyathea) in the South Island of New Zealand; they can reach 25m (82ft) in height.

an...
ev...
fo...
lig...
the...

coo...
and...
and...
pera...
are...
dom...
are...
grass...
a fe...
the...
speci...

M...

ow-
rain
nly
but

are
ests
ytic
m-
ere
rè-
ere
w—
nly
in
ew

in

dim damp places, the ideal habitat for the gametophyte generation, but some grow in and colonize open spaces and thrive in sunny conditions. In order to resist desiccation various mechanisms have evolved, the simplest being to allow the frond to shrivel as happens in some *Asplenium* species. The most common adaptation is the development of thick fleshy leaves that contain more water. However, more complex techniques are found in other species—often a barrier of scales, hairs or roots around parts of the plant, which help to retain moisture, or, in other cases possessing fronds that are easily lost under adverse conditions coupled with the ability to rapidly grow replacements when water is available again; this is frequently shown by species of *Davallia*.

Some species, e.g. *Salvinia* and *Azolla*, have overcome all water-supply problems by being adapted to an aquatic habitat; these are free-floating, forming carpets on the water surface with the roots hanging down. The fronds are small, up to about 3cm (1.2in) long, but they can grow and spread very rapidly, often blocking waterways where the water is slow flowing or still. The upper surface of the fronds is covered with specialized hairs that repel water and ensure that the plant will not sink.

Some ferns have evolved special adaptations. The species of the epiphytic genus *Platycerium*

have two distinct frond forms; the normal sterile and fertile types, and specialized basal ones which trap rotting vegetation and provide a mineral supply from the forming humus. Two other epiphytic genera, *Solanopteris* and *Lecanopteris*, form symbiotic relationships with ants which live in the swollen rhizomes.

Classification

In order to distinguish the different species of ferns, a variety of characters need to be studied, these include external structure, habit, vascular system, scales, hairs, position and structure of sori, sporangium development, spore structure and chromosome number.

Often studied with the ferns are the plants called the fern-allies. These are not true ferns, but are more nearly related to the ferns than they are to any other group of plants. They include *Equisetum* (the horsetails), *Lycopodium*, *Phylloglossum* and *Selaginella* (the clubmosses), and *Stylites* and *Isoetes* (the quillworts). The horsetails have ridged stems which in some species branch several times to give a brush-like or horsetail appearance. The leaves are reduced to papery scales which occur at the nodes, they contain no chlorophyll; the stems undertake all photosynthesis. The fertile shoots have a small cone or strobilus at the apex which contains the spores. Clubmosses have branching stems with rows of green leaves, and grow in a creeping fashion on the soil or epiphytically on trees. Sporangia are found at the base of leaves at the

apices of some growing shoots. The quillworts grow submerged in water or almost buried in moist soil and appear as bunches of plain straight green stems.

Fossil evidence shows that the ferns are a very old group of plants. The earliest remains are from the Lower Devonian period about 300 million years ago, these fragments are thought to be of an extinct group of fern-allies. True ferns did not appear for another 50 million years, until the Upper Devonian and Lower Carboniferous period. At this time many of the species were large and formed trees; with the horsetails and clubmosses they were the dominant plants and their remains form today's fossil fuels.

Ferns and man

Most fern species today have a restricted distribution, though they tend to be more widely distributed than flowering plant species. Probably the most widespread fern is bracken, *Pteridium aquilinum*: in temperate areas it is a weed often encroaching on pastureland where it is difficult to eradicate. In tropical areas one fern weed that is a problem is *Dicranopteris* which forms impenetrable thickets.

Many people consider ferns to be of little value to man other than being ornamental, but in some areas they form part of the local diet, and of more importance is evidence that chemicals found in some of them may prove useful as drugs and in insect control.

Above: The underside of a mature frond of bracken clearly shows the brown sori *in which the dust-like asexual spores are carried, to be released through slits in dry conditions.*

Top right: Three weeks after a scorching fire young bracken ferns are pushing up. They arise from growth buds of different ages on the extensive underground root system.

Bottom right: Not all ferns grow on land; a few are aquatic, like the floating Azolla, *which rapidly makes dense carpets. Special hairs on the fronds repel water so that they do not sink.*

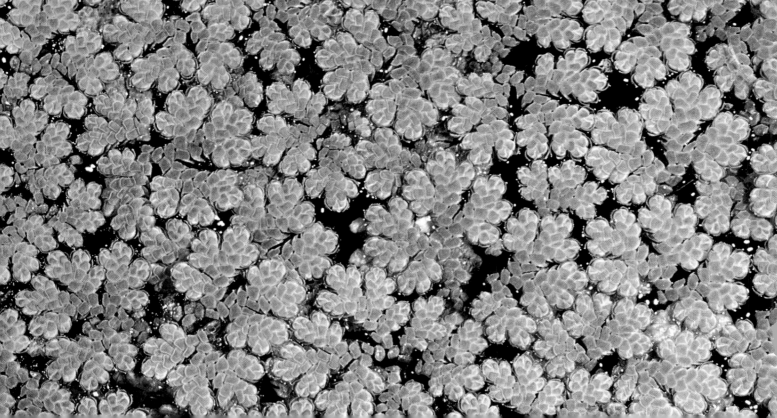

FUNGI
vegetable renegades

The fungi are unique among plants in that they are unable to make their own food by photosynthesis. Early in their evolution they lost their chlorophyll and must therefore, like animals, obtain ready-made organic food, either as saprophytes feeding on dead organic matter or as parasites.

The fine thread-like cells of fungi, called hyphae, resemble those of other plants in having cell walls and not moving. They are sometimes woven together forming solid structures as in the mushrooms and toadstools which are the 'fruit-bodies' of some species, but more usually form a loose weft, or mycelium, spreading through their food. Hyphae usually have dividing cross-walls and each cell contains protoplasm, droplets of stored oil and several nuclei.

Reproductive structures and classification

The basic unit of reproduction in fungi is the spore, a tiny parcel of protoplasm, nuclei and stored oil, sometimes less than 0.00002mm (8×10^{-7}in) in diameter, contained within a protective wall. Spores may be produced sexually following the fusion of two cells, or asexually from the vegetative hyphae. Most fungi can be identified only from their reproductive structures and are therefore classified by these.

In the lower fungi, or phycomycetes, which have hyphae without dividing cross-walls, asexual swimming zoospores with whip-like tails are sometimes produced. More often stalked sporangiophores containing many spores are formed. These burst open when ripe, the spores being dispersed by wind or on the bodies of insects. The sexual zygospores of phycomycetes are formed singly and have a thick protective wall. They may survive adverse conditions for several months, eventually germinating to form a sporangiophore liberating asexual spores.

The higher fungi are divided into the ascomycetes (the cup and flask fungi and the yeasts) and the basidiomycetes (the mushrooms and toadstools). In all higher fungi, the asexual spores are not enclosed but are produced at the end of special hyphae. These 'conidia' range in size from less than 0.00002mm to 0.1mm (8×10^{-7}in to 4×10^{-3}in), and may contain one or several cells.

In ascomycetes, the sexual ascospores are formed in groups of eight within a cylindrical ascus which shoots the ripe spores up into the wind. The fungus *Pyronema* has an ascus which bursts after absorbing water by osmosis, throwing the spores out to a distance of two to three centimetres.

The sexual basidiospores of the mushrooms and toadstools are formed uncontained in groups of four on the spore-bearing hymenium of the fruiting body. This hymenium may take the form of gills, pores or spines and is aligned at right angles to the direction of gravity so that the spores can fall out freely into the air. Consequently a toadstool laid sideways overnight in a collector's basket can be found in the morning to have turned its cap horizontally. The spores are liberated in an ingenious way: each ripe spore has an attendant drop of water which, as it dries, sets up surface tensions which cause the spore to be shot off the hymenium.

The Fungi Imperfecti is a class with no known sexual reproduction and their classification is therefore uncertain. Spores are produced in huge quantities; the bracket fungus *Ganoderma applanatum* may produce as many as 30 billion a day, and 5,500 billion in a year. A common mushroom, *Agaricus campestris*, measuring 10cm (4in) across can produce 1.6 billion spores in the first few days of its existence, whilst the American giant puffball *Calvatia gigantea* has fruiting bodies which may reach 1.7m (5.5ft) in diameter with correspondingly huge amounts of spores.

To germinate, a spore must have moisture, warmth and suitable food. Many spores are wasted but the vast number produced ensures the continuation of a species. Many spores are found high in the atmosphere and may be carried long distances in the jet stream, which helps explain the worldwide distribution of many species, such as *Aspergillus niger* which can be found from the arctic to the tropics. In fact, their effective dispersal combined with fast growth, great variety and ability to evolve and adapt quickly make the fungi a ubiquitously successful group, far more common and varied than the sight of the occasional toadstool might lead us to believe.

Right: The sulphur tuft fungus, **Hypholoma fasciculare,** *always grows on dead wood and is frequently seen at the base of dead trees or on cut trunks, making large clumps.*

Amanita muscaria
Fly Agaric

Feeding and growth

Fungi employ several methods of obtaining the organic food they require. Most species are saprophytes, feeding on dead plants and animals. Some, such as the black mould *Aspergillus* and the green mould *Penicillium* thrive on a wide variety of substrates from old books to dead leaves; others are less catholic. With the bacteria, these fungi dispose of the world's waste; the all-too-obvious building-up of plastic wastes in our environment is a testament to the efficiency with which other digestible organic wastes are dealt with by these agents of decay.

In nature's cycle of growth and decay, the fungi are essential in releasing from their food organic salts needed for plant growth. The growth of the fairy ring champignon *Marasmius oreades* graphically illustrates the results of decay by fungi. Having begun from a single spore, the mycelium of the fungus advances outwards, at a rate of 15–30cm (6–12in) a year, forming a circular patch. At the outer edge of the ring fungal decay of organic matter increases the mineral salt content of the soil, particularly of nitrates, causing luxuriant grass growth. Inside this is a ring of poor growth where the thriving mycelium clogs the soil reducing aeration and drainage and stunting plant growth. It is here that the fruiting bodies appear. Towards the centre, where the ageing mycelium is dying, releasing minerals to the soil by bacterial decay, is another ring of rich growth.

Saprophytic fungi

Essential as they are to the fertility of our soil, some saprophytic fungi cause great damage to man economically. Many fungi are capable of digesting the resistant lignin and cellulose in wood, as is obvious from the proliferation of bracket and cup fungi on both living trees and dead wood. The dry rot fungus *Merulius lachrymans* partly avoids the need of most species for damp conditions by transporting water in its long spreading root-like rhizomorphs which are capable of penetrating bricks and stone. Storage rots of grains, fruits and vegetables also cause considerable damage to man's crops. Species of *Rhizopus* cause soft rot in potatoes, peaches, tomatoes and other crops. *Fusarium caeruleum* causes dry rot in potatoes, and various species of *Aspergillus* cause deterioration in grains, sometimes producing poisons too, such as the aflotoxin which killed 10,000 turkeys fed on groundnut meal imported into Britain in 1960.

Of great use to man in brewing and baking are the yeasts, a uniquely one-celled group of ascomycetes reproducing by asexual budding. Other fungi hold great promise for future food production. *Aspergillus niger* is already used in the industrial production of citric acid and the making of soy sauce.

Fungal parasites of plants

Many fungi are parasitic, obtaining their food

*Above: stinkhorn, **Phallus impudicus**, produces its spores in dark green mucus, with a disgusting stench which attracts flies. These eat the mucus and spread the spores in their excreta.*

*Top right: The minute spores of puff balls like **Lycoperdon perlatum** are released through a small opening at the top. Here a rain drop landing on the fungus makes it expel a spore cloud.*

Right: The mould on this rotting apple is a fungus. Many microscopic fungi grow on decaying organic substances, while others attack live plants and cause diseases. Penicillin is derived from a mould.

from living organisms, usually plants. Most are facultative parasites being capable also of living saprophytically. *Pythium*, for example, which causes damping-off in seedlings, usually lives saprophytically in the soil. It has a very wide host specificity, attacking a variety of vulnerable seedlings when the opportunity arises. Obligate parasites such as the black rust of wheat *Puccinea graminis tritici* can only exist as parasites and are often host-specific, parasitizing only one species. Plant parasites have caused immense damage to crops since the plagues of biblical times, and today demand the expenditure of millions of pounds a year on fungicides and preventative measures. Monoculture, the

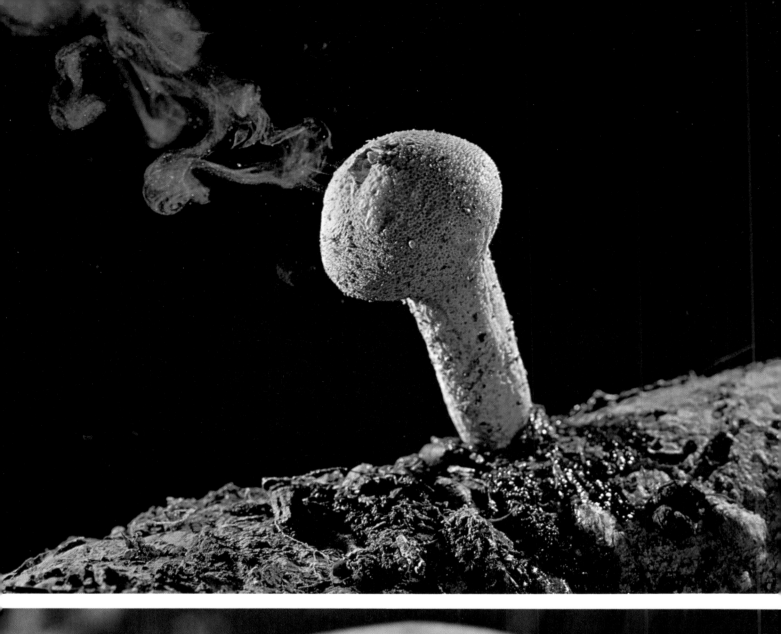

growing of one crop over large areas, renders crops particularly vulnerable. The great Irish famine of 1845–60 was caused by the potato late blight fungus, *Phytophthora infestans,* which forced 1.5 million Irish people to emigrate to the USA and starved another million to death.

Many plants are attacked by more than one fungal parasite. Besides late blight, the potato suffers early blight, *Alternaria solani;* black scurf, *Rhizoctania solani;* common scab, *Streptomyces scabies;* powdery scab, *Spongospora Subterranea* dry rot, *Fusarium caeruleum;* and several others attacking through the plant's life cycle.

The vine *Vitis vinifera* also suffers several fungal diseases. Among them is downy mildew produced by *Plasmopara viticola,* a fungus originally native to America. Around 1885, the aphid *Phylloxera* was accidentally introduced into France from America, decimating the economically essential vineyards. To combat this, resistant root stocks were imported. This defeated *Phylloxera,* but introduced downy mildew, which again destroyed vast areas of crops. The French scientist Alexis Millardet noticed that vines near paths sprayed with a mixture of copper sulphate and lime—to deter hungry passers-by—were much healthier. Working from this he developed the first fungicidal spray, which saved the French wine industry, and is used to this day under the name 'Bordeaux mixture'.

Cereals form a major part of the world's food production, and the smuts and rusts that cause such havoc among them are of enormous economic importance. Black rust of wheat parasitizes two hosts during its life cycle. In autumn, resistant overwintering spores are produced, and these germinate in the spring to infect not wheat but the shrub, barberry. After a short period parasitizing barberry, spores are produced which infect the now thriving wheat. Spores produced on the wheat spread the disease rapidly throughout the crop until autumn when overwintering spores are produced again. In America, a campaign to exterminate the barberry and thus cut the rust's life cycle has had some success in the north, but not in the south where the fungus can miss out this step. The breeding of resistant strains has been more successful, but the rapid evolution of the parasite and the large number of rust strains in existence make this a winning battle between parasite and plant breeders.

The ascomycete ergot, *Claviceps purpurea,* infects the ovaries of rye, reducing them to a mass of hyphae producing a sticky insect-attracting honey-dew. Ergot is extremely poisonous, causing a painful death to man and grazing animals, and serious outbreaks of ergotism, or St Antony's fire, caused by eating contaminated grain, occurred as recently as 1951 in Pont-St Esprit in France. A muscle-contractant used to control haemorrhaging during childbirth is extracted from ergot, perhaps explaining why farmers have long avoided certain fields where many stillbirths

Right: Cap-forming fungi (mushrooms and toadstools) usually have either gills (above) or tubes (below). Within these spores are formed. There is a very small space between gills, into which the spores are literally shot off the sides to a distance of 0.1mm (0.004in). They then fall downwards and are carried away on any air currents. A mushroom with a 10cm (4in) cap produces around 16 thousand million spores.

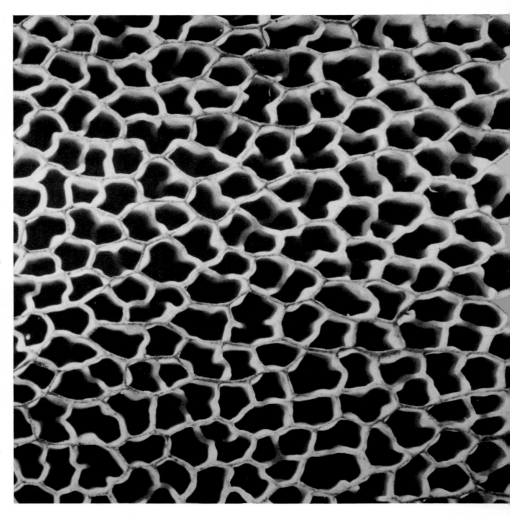

Mushrooms under control

The common mushroom, *Agaricus hortensis*, is one of the few larger fungi which have successfully been taken into cultivation. This began in France in the 17th century, following the use of hot-beds of horse-manure which were often found to sprout mushrooms. Commercial mushroom cultivation today is a complicated and very precise operation carried out on a large scale in light-proof sheds or sometimes under-ground. The rich, manure-based compost used is 'spawned' with small lumps made up of the 'roots' or hyphae of the mushroom.

have occurred among their animals.

Fungal parasites may annihilate their host species; the American chestnut has been completely wiped out by *Endothia parasitica,* and Dutch elm disease, caused by *Ceratocystis ulmi* and carried by wood-boring beetles, threatens the elm with the same fate.

Fungal parasites of animals

Fungal diseases of animals are less common. The fly fungus *Entomophthora muscae* is one of several fungi which parasitize insects. The mycelium, growing from a single spore landing on the host, grows through the fly's body, eventually leaving an empty husk. These victims are often seen clinging to dusty window panes, surrounded by a halo of emerging spore-bearing hyphae. Another group, the hyphomycetes, feed on eelworms, trapping them in looped sticky hyphae.

In man (and other mammals) fungi such as *Epidermophyton floccosum* produce the skin diseases athlete's foot and ringworm, which may reach epidemic proportions. Others, such as *Cercospora apii,* can cause severe disfigurement. Aspergillosis, or farmer's lung, is caused by the common *Aspergillus niger,* often found on fruit. Several other fungi produce severe lung diseases such as blastomycosis caused by *Blastomyces dermatitidis* in America, often misdiagnosed as tuberculosis. Many of these fungi are normally soil-dwelling saprophytes, and soil may be an important source of infection.

Symbiotic fungi

Some fungi form nutritional relationships with other plants which are mutually beneficial to both parties. Such a relationship is called a symbiosis. In lichens the fungus gains nutrients by parasitizing living algal cells and feeding on dead ones, whilst the algal cells are protected from high light intensities and dehydration. Mycorrhiza, associations between fungi and the roots of higher plants, usually trees, are very common, especially with conifers. A thin layer of mycelium covers the roots, sometimes invading the root cells, and the tree seems to benefit from the efficiency of the fungi in absorbing

Amanita mappa
False Death Cap

water and mineral salts from the soil, while it is thought that the fungus gains sugar from the roots. Some mycorrhizal associations are specific, *Boletus elegans* for instance only grows with the larch *(Larix decidua)*.

Other fungi produce mycorrhiza with many tree species. The dependence of the fungus on the tree varies, but some seem only to grow in these associations. Members of the *Ericaceae,* the heather family, seem particularly dependent on mycorrhiza, some species never being found without them. The fungus may even be found in the seeds, as with *Phoma radicis.* In the orchids, the protocorm which grows from the tiny seed, will not develop until infected by the appropriate mycorrhizal fungus from the soil, which may take two years.

In most symbiotic relationships, the partners are 'fair weather friends'. If the delicate balance between parasitism and symbiosis is tipped due to conditions favouring one or the other partner, the relationship may cease to be mutually beneficial and break down, causing the parasitic destruction of the host.

Survival mechanisms

The fungi have evolved several very effective survival mechanisms. Many toadstools are distasteful and some are actually poisonous, ensuring that the minimum number are eaten and the maximum number survive to produce spores and propagate the species. The collector avoiding the Death Cap *(Amanita phalloides)* ensures its spread. Several saprophytic fungi such as *Penicillium* and *Streptomyces* compete directly with bacteria for food and, by producing antibiotics to kill them, eliminate the competition and thus thrive.

CONIFEROUS TREES
trees with cones and needles

Conifers are a peculiar, but geographically and economically important, group of trees that ripen their seeds on the scales of woody cones. Two other common terms, coniferous and cone-bearing, mean just the same thing.

Conifers are also called gymnosperms, from the Greek meaning 'naked-seeded'. This applies to the position of the seed, and the ovule from which it develops, on the surface of the cone scale. It is not enclosed within an ovary, as in broadleaved trees and flowering plants, which are called angiosperms, or 'hidden-seeded'. As the fertilized conifer seed develops, it becomes tightly enclosed between the cone scales, which protect it from birds, mammals and insects. Finally, when the cones ripen and the scales open, it is exposed again and eventually falls out.

Conifers are also called softwoods, because the timber of most species is distinctly softer than that of most deciduous trees, or hardwoods. It is therefore easier to work. and under modern industrial conditions this makes it more valuable for all the everyday uses to which timber can be put, such as house-building, packaging, fencing and the manufacture of chipboard and plywood. Softwoods have long fibres, which makes them highly suitable for making paper—basically an interlocked sheet of wood fibres. Naturally some kinds of conifers are better for each job than others. But softwoods are adaptable materials, and in practice the industrialist adjusts his process to match his source of supply.

The leaves of nearly all conifers are narrow and needle-shaped, and therefore the group as a whole are often called needle-leaved trees. This leaf structure is better than that of broadleaved trees for restricting the loss of water into the air through transpiration. Therefore conifers can thrive in drier places, like the subtropics of Spain, South Africa or California, than can broadleaved trees. Because cold soils reduce the water supplies available for plant life, conifers also thrive farther north, and higher up mountains, than do deciduous trees. They also need smaller supplies of mineral nutrients, so grow faster on poorer soils than deciduous trees can do. Foresters, the world over, naturally tend to make new plantations with profitable conifers, rather than with the deciduous trees that demand better soils for slower timber outputs.

Conifers are often known as evergreens. Nearly all conifers have evergreen leaves, which live for three or four years before falling. Therefore their branches are never leafless. There are two common exceptions—the larches, *Larix* species, which grow high on the mountains, and the swamp cypress, *Taxodium distichum*, which lives under difficult soil-water conditions in the swamps of the southern USA. The evergreen habit is efficient, because it means that photosynthesis can be carried on, whenever the temperature is high enough, all the winter through.

Pinus sylvestris
Scots Pine

Another peculiarity of the conifers is the presence of resin—they are sometimes called resin-bearing trees. Resin is a transparent sticky fluid, composed of a solid wax, called rosin, dissolved in a lighter spirit, known as turpentine. It is formed in channels or small chambers in wood, bark, needle-shaped leaves, cones and even seeds, and is easily detected by its sharp distinctive smell. If the tree's tissues are injured, resin oozes over the wound. The vola-

tile turpentine evaporates and a protective coating of hard rosin remains. This shields the tree—not always successfully—from attacks by fungi, insects, mammals and birds, as well as man-made injuries.

Resin is harvested commercially in south-west France and the southern USA, where it is called 'naval stores', by tapping living pine trees. The turpentine is distilled off for use as a solvent for paints and chemicals. Most of the rosin goes into printing inks; it gives them quick-drying properties essential for high-speed machine printing.

Simple woody stems

The wood of conifers is simpler than that of deciduous trees. It has no pores or vessels—those special structures for rapid transport of sap in spring that the deciduous tree needs when

Above: Huge areas of virgin conifer forest still remain virtually untouched in parts of North America, like that of alpine fir (Abies lasiocarpa) in the Bow Valley in British Columbia.

its broad-bladed leaves expand. The conifer's need for more sap in spring is met more simply by developing broad bands of springwood, composed of long thin-walled fibrous cells with large cavities available for sap-flow. In summer the tree's need for greater support for its enlarged crown is met by a change to summerwood, composed of similar cells with thicker walls and smaller cavities. Many conifers, especially pines, *Pinus* species, and Douglas firs, *Pseudotsuga*, develop very clear annual rings in this way. In others, such as spruces, *Picea*, the summerwood is less obvious.

The thickening of conifer stems is due, like that of the woody stems of deciduous trees, to the activity of a layer of cambium cells on the outer surface of the true wood. This cambium layer lies between the sapwood, which carries the main sap stream upwards from the roots, and the bast, which carries the downward flow of nutritious sugar-sap from the leaves to the lower stem and underground root tissues.

Conifer bark, which is formed as a protective layer, outside the bast, by its own cork cambium, shows marked variation according to kind of tree. Some species, like the western red cedar, *Thuja plicata*, have thin stringy bark. Others, like the coast redwood of California, *Sequoia sempervirens*, bear extremely thick fibrous bark, which can even protect the living stem against the heat of a forest fire.

As conifer stems grow larger, their inner zone of wood ceases to carry sap, as sapwood. It is then converted to heartwood, which serves only to support the tree's crown. This change is effected by the deposition of resin and related chemicals. With some kinds, there is an obvious change in colour and durability. Larches, *Larix* species, develop red–brown heartwood that resists decay much longer than does the pale-brown sapwood. But in spruces, *Picea* species, the heartwood is scarcely darker, and certainly no more durable, than the sapwood.

Conifer stems also hold ray tissues, which radiate from the centre, or from points near it. They carry nutrients sideways, and store food reserves through the winter, but they are never so thick or obvious as those of some deciduous trees.

Life pattern

Conifers start life as seeds, small brown oval grains that fall from the cones of their parent trees, attached to thin triangular or oval wings. These wings aid their spread, enabling the wind to carry them a fair distance from the parent tree before they strike the ground. With the warmth of spring, the seed takes in moisture from the damp earth, swells and grows a little root. Once this root has become anchored and can draw in water through its root hairs, it raises the rest of the seed upwards. It bends its stalk until it stands erect. The seed coat or husk then splits and falls away, and the seed-leaves within it expand. Many conifers, unlike other plants, have numerous seed-leaves; common numbers are two for western red cedar, *Thuja*, three for hemlock, *Tsuga*, but twenty or so for pines, *Pinus* species. These seed-leaves quickly turn green and begin the photosynthesis that nourishes the seedling tree, once the food reserves in the seed have run out.

A leading shoot soon emerges from the centre of the whorl of seed-leaves and bears green needles. These early juvenile needles are often different, in arrangement or character, from the adult needles that the tree will bear later. Pines, for example, bear grass-like solitary juvenile needles, but their adult needles are always grouped in twos, threes or fives. After its first

summer the seedling develops a main resting bud at its tip, with smaller side resting buds around it, and growth ceases for the winter.

Next spring these buds burst. Their scales fall away, and the shoots within them make remarkably rapid growth for a spell of a few weeks. The terminal shoot goes upwards, to become a leader. Side shoots go outwards, to become the tree's first branches. As they extend, the needles that were already hidden in the bud expand all around them. After this rapid expansion a pause ensues. By midsummer growth has usually ceased and the winter buds, that will expand in the following spring, are already being formed.

Sudden spurts of growth on a geometrical pattern, are characteristic of conifers. The age of most young, and indeed middle-aged, trees can be ascertained by counting the numbers of

annual growth whorls, or clusters of branches, up their stems. The length of the unbranched shoot, or internode, between successive nodes of branches, shows how far the tree grew taller each year; older trees, however, lose this regular pattern.

Conifers, overall, do not grow much larger or faster, than deciduous trees, nor do they live much longer. But the wood they produce is concentrated in a convenient form in straight lightly-branched main trunks, rather than the heavy branching crowns of the deciduous trees. Further, the conifers can be grown closer together. This makes them more profitable to the forester and more attractive to the timber merchant.

Record-breaking conifers exceed the largest and oldest deciduous trees. Coast redwoods, *Sequoia sempervirens*, the world's tallest trees

Above: Most conifers bear their seeds within woody cones like these of wellingtonia (Sequoiadendron giganteum); thick woody scales protect the seeds till they are ripe.

Far left: The resin of conifers has been used for many purposes since ancient times. Here a maritime pine (Pinus pinaster) is being tapped for turpentine production in Portugal.

Left: New needles of Scots pine, Pinus sylvestris, breaking through the brown bud scales which have been protecting them as they develop.

today, soar to 117m (384ft) in California. Douglas firs felled at the end of the last century almost certainly exceeded 122m (400ft). Their allies, the giant sequoias, *Sequoiadendron giganteum*, also Californian, can reach circumferences up to 30m (98ft) measured 1.2m (4ft) above ground level. They show the greatest timber volumes of any known tree. Bristlecone pines, *Pinus aristata*, in the Arizona desert, may live for 4,000 years, remaining stunted trees in a harsh environment.

Reproduction

Conifers that grow in the open and receive full sunlight, often start to flower when only 15 years old. Those crowded in plantations may postpone flowering to an age of 30 years or more. Once flowering starts, it continues annually, with occasional years of poor seed production. During its lifetime a big spruce or pine may bear a million seeds, in a lavish provision to ensure the survival of its kind.

All conifer flowers are borne in separate male and female clusters. Each individual tree is capable of bearing flowers of both sexes, though it does not always do so in the same year. Pollination is always effected by wind, possibly because conifers evolved when there were no insects to carry pollen. Consequently their flowers have no green sepals, no brightly colour-ed petals, no nectar and no attractive scent. A few—notably the female blossoms of larches and the Korean silver fir, *Abies koreana*—are very pretty, but this is just an accident! Nearly all conifers flower in spring. The true cedars of the *Cedrus* genus are an exception; they flower in autumn.

Male conifer flowers open in clusters, oval or cylindrical in shape like the male catkins of deciduous trees. Each flower consists of small bracts and yellow anthers, which shed abundant golden pollen. For a week or two the flower cluster looks like a golden cone or caterpillar; then it fades. Most female conifer flowers are bud-shaped, green and inconspicuous; they open at or near the tips of branches. Pines bear round red flowers, each smaller than a pea, at the tips of new shoots. Larches have larger ones, red or white, shaped like rosettes. Spruces and silver firs bear small upright green conelets. Each female flower consists of scales, set spirally on a central stalk. On the inner face of each scale lie the ovules which, after fertilization by male pollen, become the seeds. In most conifers, each scale bears two ovules, but cypresses, *Chamaecyparis*, have four or five, and monkey puzzle, *Araucaria*, only one.

After fertilization, the female flowers enlarge and develop through a soft green stage into hard woody cones. With most kinds, this process takes about six months. Pines need 18 months; after 6 months one female flower has only grown into a brown pea-sized structure, which needs another year to mature. Cedars, *Cedrus*, take two or three years. Cones vary markedly in size and shape, and provide a handy means of identifying each kind of tree. In a typical cone, each scale is attached to the central stalk by a hinged joint. When the weather gets drier, one side of this joint contracts, the angle of the scale alters and the cone opens. When the air gets moister, the scales close again, and this explains why ripe cones can be used, in a rough and ready way, to forecast the weather.

Ripe seeds fall gradually, day by day, and usually during dry spring weather, from the open cones. Each has a papery wing, which delays its fall as it travels on the wind to some point away from its parent tree. There, as soon as rain falls, it takes in moisture and sprouts, and the whole life cycle begins anew.

Conifer seeds have many enemies. They provide tasty nutritious food for woodmice and squirrels. In America, chipmunks regularly collect cones and store them in hoards called caches, as a food reserve for the winter. With his needle-sharp teeth a squirrel can easily bite through a cone scale to reach the tasty seed below. Birds gobble up fallen seeds. One kind of finch called the crossbill, *Loxia curvirostra*, has its two mandibles crossed; this enables it to prize back cone scales to reach the seed below. Only a small percentage of seeds survive to germinate, and the seedlings only thrive if they start growth on bare, or nearly bare, soil. They grow very slowly for their first few years and are easily choked by other plants. Some young conifers can thrive under partial shade, but others, in-cluding pines and larches, need full sunlight for successful growth.

Right: The boles of Douglas firs, **Pseudotsuga menziesii,** *on Vancouver Island, B.C. These conifers are extremely important timber trees and have been widely planted in temperate regions. They grow very fast and combine great height with relative slimness: a specimen 100m (330ft) tall may only be about 3m (10ft) wide at the base. A Douglas fir cut down in 1895 was 129m (426ft) tall, and thus probably the tallest tree ever recorded, although the record is not scientifically authenticated.*

Below: The first sight of the female cone of Scots pine, emerging with new needles in May, still soft and minute (much magnified). This cone will eventually become large and woody.

Coniferous forests

The world distribution of coniferous forests comprises a broad band that reaches right round the Northern Hemisphere, across Canada and the northern United States, then across Europe from Scotland to Russia, and finally across Siberia to the Pacific Ocean. Farther south there are great conifer forests spreading along mountain chains, including the Rocky Mountains and Appalachians in North America; the Alps, Pyrenees and Apennines in Europe; the Caucasus; the Himalayas in Asia; and the higher hills of Japan. A few conifers grow on tropical mountains, and a very few, like the monkey puzzle of the Andes, in the Southern Hemisphere. The general pattern is that the conifers grow in colder places than deciduous trees, though some, like the maritime pine, *Pinus pinaster*, of the Mediterranean region, thrive under climates that are too hot and dry for deciduous trees to tolerate. In general, the conifers are able to flourish on the poorer soils, too infertile to attract clearance for agriculture, and also on steep hill slopes unsuited to cultivation.

These coniferous forests form an enormous, and fortunately renewable, reserve of raw material for industry. Wherever the woods can be reached by modern transport, the loggers go in and harvest the timber. The larger logs are sawn into planks and other structural timber for house-building, packaging, fencing and everyday furniture, or are peeled to provide flat sheets of veneer for making plywood. Smaller logs are used as telegraph poles, mine props and fence posts. Some are cut into chips, which are then stuck together with plastic glues to become chipboard, a 'man-made' wood. Very large quantities are ground or cooked with chemicals to make paper pulp, the source of all our newsprint, writing, book-printing and wrapping paper. Nowadays the owners of these forests, who may be governments, timber or paper companies, or individual landowners, plan this harvest for many years ahead. It is only worthwhile to build a big modern sawmill or paper mill if it is known that timber can be cut to feed it, far into the future. Ideally, the amount of timber harvested each year should equal the amount that grows again in the forests within economical transport distance.

Pinewoods
Conifers of many kinds are loosely called 'pines' or 'firs', but all the true pines belong to the genus *Pinus* and are identified at once by their slender needles grouped along the twigs in twos, threes or fives. Each needle is tough and leathery, and has bent surfaces, concave within and convex without. In hot dry weather it contracts and becomes almost cylindrical in cross-section. This lessens water loss through transpiration and enables pines to grow in hot dry places. There are many species, each adapted to peculiar climates, from the tropics to the far north. The long-leaf pine, *P. palustris*, of the southern USA, has needles up to 30cm (12in) long. The Scots pine, *P. sylvestris*, of northern Europe, and the Jack pine, *p. banksiana*, of northern Canada, bear very short ones, only 5cm (2in) in length. Some pines have small cones; others, like the

Left: A South African cycad, Encephalartos villosus. These are very ancient plants, hardly changed since they first appeared 200 million years ago, which are related to conifers.

big-cone pine, *P. coulteri*, of California, ripen enormous ones, the size of pineapples. Pines cast only light shade and the ground beneath them is usually covered with short grasses or low shrubby plants like heather and bilberries. They are the haunt of red and roe deer, agile squirrels, black grouse and a giant grouse called the capercaillie.

Sprucewoods
Spruce trees can be identified by the little woody pegs set at the base of each needle, which is always solitary, never one of a group. The needles down opposite sides of the twig form a two-ranked herringbone pattern, lying in one flat plane. The cones are long, cylindrical, and always hang down when ripe. Spruces grow typically on wet swampy land all round the world's northlands, and on mountainsides with ample rainfall farther south.

Spruce trees tolerate deep shade, and can crowd close together, forming dense woodland. They cast an even deeper shade. Hardly any light reaches the floor of most sprucewoods, and no green plants can live in such places. The ground is covered with dead needles and twigs that break down slowly to form an acid mould or humus. But the nutrients this holds gradually return, via the roots, to the trees, and spruce forests yield high volumes of useful timber in perpetuity. Birds and mammals are scarce; pigeons nest and deer hide among the spruces, but feed elsewhere.

Larchwoods
Larches are easily recognized by the grouping of most of their needles in clusters on little knobs along their twigs; at the twig tips, however, the needles are borne singly. These needles fall each autumn, but all through winter the woody knobs on leafless twigs make identification easy. The cones are barrel-shaped. Larch needles are thinner and softer than those of other conifers, bright green in spring, they turn bright orange when they fade in autumn. They cast only a light shade, and soft grasses, ferns and some flowering plants flourish beneath them. Natural larchwoods are characteristic of high mountain ranges, where they provide shelter and pasture for sheep and cattle, as well as homes for wild deer. Larch timber is exceptionally tough, so is used for exacting work like ship's planking, bridge-building and beams and rafters in wooden buildings.

Quaint survivors from the prehistoric past
Linked with the true conifers are several groups of primitive trees which survive in relatively small numbers of species and rarely form substantial forests. Best known are the yews, *Taxus* species, which have needle-shaped evergreen leaves, but bear their seeds in fleshy pink berries, not woody cones. Every yew tree is either male or female; male flowers open as yellow catkins in spring; female ones are green and bud-shaped. The rare maidenhair tree, *Ginkgo biloba*, from China and Japan, bears fan-shaped leaves like the fronds of a maidenhair fern; they spring from woody knobs and fall each autumn. Maidenhair's male flowers are catkin-like; the small female flowers, on separate trees, ripen plum-like fruits with hard stone-shaped seeds. The cycads, such as *Cycas revoluta*, are palm-like trees that grow in the tropics and expand huge feathery leaves arranged spirally on an unbranched stem. They bear male and female flowers in separate cone-shaped structures, and ripen large seeds resembling plums, with a fleshy outer layer and a hard stone within. Their leaves are used for the thatched roofs and walls of huts, but the wood has no value.

DECIDUOUS TREES
trees that lose their leaves

Deciduous trees shed their leaves at the approach of winter, or in some countries that have hot rainless seasons at the advent of the annual drought. They are also called broadleaved trees because nearly all of them bear leaves with broad thin blades that contrast with the narrow needles of the conifers. Another common name is hardwoods, because the wood of most, though not all, is distinctly harder than that of coniferous trees, which are therefore called softwoods. Yet another name demands explanation, and that is evergreen. If a tree, whether broadleaved or coniferous, never sheds all its leaves at once, it is called an evergreen, even though individual leaves eventually drop off after a life-span of two, three or more years.

Seasonal leaf pattern

The distinctive feature of typical deciduous trees is their annual rhythm of life and growth. They stand leafless through the winter. When the warmth of spring stirs them, their remarkable winter resting buds burst. Scales fall away and release delicate shoots that bear green leaves of a shape peculiar to each kind. These leaves function actively all through summer, carrying out photosynthesis in just the same way as the leaves of smaller plants. When autumn arrives, the green colouring matter called chlorophyll, that is essential to this process, breaks down. The leaves change colour to yellow, orange or scarlet, due to pigments called carotins, or else

Aesculus hippocastanum
Horse Chestnut

An ancient stag-headed oak tree in the New Forest photographed in May (inset), when the fresh green leaves have just appeared, and in winter when all have fallen.

to purple hues caused by other pigments known as xanthocyanins. At the same time the food reserves of the leaves are withdrawn into the tree's woody stem. After a brief display of brilliant colour, each leaf is cut off from the twig that bears it by a thin layer of cork, which forms across the base of its stalk. It then falls, usually on a windy day, or after a sharp night frost has caused ice to form at the point of attachment; the morning thaw snaps the link.

Temperature is the controlling factor that causes these leaves to be shed, but it works in a curious way. The leaves can only function whilst they have an ample supply of sap sent up by the roots, but the roots in turn can only draw water from the earth when the soil temperature is over 4°C (39°F). If an ordinary broad-leaved tree retained its leaves in the winter, it would quickly lose more water than it could replace, and it would die of drought. Evergreen broadleaved trees have special devices, including waxy surfaces and small breathing pores, to limit transpiration and these enable them to survive. But they rarely grow so fast as the true deciduous or 'summergreen' trees, because the thin blades of the latter are very efficient mechanisms for using the sun's energy for growth. So efficient indeed that these trees can afford to grow a fresh set of leaves each spring and discard them each autumn. In the forest, however, the spent leaves enrich the leaf mould at ground level. The nutrients they hold, chiefly mineral salts, are picked up by the tree's roots and re-cycled into the tree's crown, so they are not lost for ever.

The woody stem

The leaves of deciduous trees are supported on the woody twigs and branches of a sturdy trunk, which increases in thickness as the tree grows older. All these woody stems, large and small, have an unseen but vital pattern of seasonal growth that keeps in step with the leaf life rhythm. In spring, when the fresh leaves are opening and demanding large supplies of sap, a ring—more strictly, a cylinder—of cells, called the cambium, forms a fresh band of open-textured cells and larger conductive elements, known as vessels, to transport it. This is the springwood. In the summer, when more support is needed for the enlarged crown of foliage, it adds closer-textured material, with thicker cell walls and less pore space, called the summerwood. The two bands of tissue, taken together, form one annual ring, and the last-formed ring is always farthest out. In this way, each trunk and branch is built up of successive layers of wood, each encircling those created earlier.

So long as the wood is actively carrying root-sap upwards, it is called sapwood. As stems become older and larger, a central core of wood is gradually transformed into heartwood, which ceases to carry sap and serves only for support. Heartwood is usually darker than sapwood, and owing to chemical changes it becomes, in some species only, more durable. Oak, for example, has very durable heartwood, but that of beech or ash is 'perishable'—it will not last if used in damp surroundings. Rays are another essential feature of woody stems. These are narrow bands of tissue that radiate out from the centre of each stem, at right angles to the annual rings. They transport nutrients sideways rather than up or down, and so store them through the winter, when photosynthesis is suspended by cold and leaf-fall.

Outside all this hard wood lies another circle, or cylinder, of softer tissue called bast. This thin, but vital layer carries sugar sap down from the leaves, where it is manufactured, to the lower stem and roots, which need it for life and growth. The true bark, which lies outside both bast and wood, protects everything within it from drying out, from casual injury and from sudden changes of temperature due to sharp frosts or blazing sunshine.

Woody stems, like the shoots of other plants, increase in length through the work of growth points, meristems, at their tips. After their first season the tissues of the soft young shoot's veins link up to form a continuous ring, or more strictly a narrow cylinder, of cambium cells, so creating a growth layer. By creating new wood on its inner side, and more bast cells on its outer side, the cambium causes the stem to grow stouter but only the meristem at its tip can make it grow longer. The highest bud at the tip of a tree usually grows more vigorously than the others, causing the main trunk to grow taller than side branches.

The wood of deciduous trees is a more complex substance than that of coniferous trees. It holds pores or vessels, composed of clusters of cells, that permit rapid upward transport of sap; they are never found in conifers. There are marked differences between timbers of various kinds of deciduous trees, which man makes use of. Ash, for example, gives tough springy tool handles; but elm is preferred for chair seats, because it will not split when the legs are driven in.

Life pattern

The deciduous tree that bears these leaves and builds up this woody stem starts life as a little seedling, which opens seed-leaves, always two in number, when it sprouts. Juvenile leaves, often simpler in outline than the familiar adult leaves, next appear on a slender upright shoot. After its first summer, the seedling sheds all these little leaves and forms resting buds, one at its tip and others at the sides. Next spring, it expands more shoots, bears more leaves and thickens its stem. This chain of processes, repeated yearly, forms first a sapling, then a large adult tree. Few deciduous trees exceed 36m (120ft) in height or 5.5m (18ft) in circumference, which is measured 1.2m (4ft) above their base. Exceptional specimens, such as a lime tree at Duncombe Park in Yorkshire, approach 45m (150ft) in height. A few old oaks exceed 12m (40ft) in circumference. The average lifespan of a deciduous tree is around 150 years, but counts of annual rings show that some old oaks may stand for 500 years or more.

If a tree escapes being felled for timber or blown down by a gale, it eventually falls victim to decay caused by fungus. Fungal spores, spread by the wind to some surface no longer protected by bark—such as the stub of a broken branch—germinate and give rise to hyphae or fungal threads that spread through the tree's inner timber, and feed on it destructively. Some fungi kill trees by blocking the phloem vessels, as in Dutch elm disease. After some years have elapsed, the trunk collapses. Before this happens the fungus has usually borne sporophores—shaped like mushrooms or brackets, that spread further spores on the wind, to perpetuate

Right: One of the trees most prized in gardens for its autumn leaf colour is the North American tupelo, Nyssa sylvatica, one of the trees that make 'fall' in New England so spectacular.

its race. The appearance of these sporophores is often the first sign of hidden decay.

Reproduction

Most deciduous trees do not flower until they are 20 years old or so. By then they have grown fairly tall and have expanded their crowns of branches and leaves. Some, like birch, then flower annually. Others, such as oak and beech, flower in some years but not in others. Flowering, when it occurs, is always abundant. A birch may bear 100,000 seeds each year for 50 years, or 5 million seeds in all, though only one seed is needed for its eventual replacement!

The flowers of deciduous trees fall into two distinct groups. Some, like cherry and horse chestnut, bear typical 'perfect' or hermaphrodite flowers. These have rings of green sepals to protect them in bud, and bright petals, either white or gaily coloured, to attract pollinating insects. Within these rings stand stamens with golden anthers that dust the visiting insect with pollen, and a pistil which receives pollen carried from another flower. Nectaries at the base of the blossom secrete nectar to reward the pollinating bee, or similar insect. Perfumes attract it.

In contrast, many other trees bear wind-pollinated flowers that lack all such attractions—they have no petals, scent or nectar. As a rule the male stamens are borne in a separate structure from the female pistils. Sometimes, as with poplars and willows, males and females are found on separate trees. Instead of sepals, we find small leafy blades called bracts that protect the sexual elements in bud. This kind of inflorescence, or flower-cluster, is called a catkin, though most are very different from the 'pussy-willow' catkins that originated this name, which means 'kitten'. Male catkins often hang down in long 'lamb's tail' bunches yellow with golden pollen. Female catkins, usually very different in appearance, may resemble buds, as in hazel, or cones, as with alder. Most catkins open before the leaves do. This makes it easier for the pollen to spread on the wind from male to female flowers.

Deciduous tree seeds, whether they develop from perfect or catkin-type flowers, show a surprising variety of size and form. Some are fleshy nuts, tiny in hornbeam but large in beech, or as the acorns of oaks. These depend for dispersal on the birds and squirrels that eat a large proportion, but drop and forget a few. Others are hard stones within fleshy fruits that attract birds. The cherry is a good example. Others again have wings to aid dispersal by wind, as in the maples and the birches, or tufts of hairs that act as parachutes, as with the willows.

The larger seeds produce big seedlings with strong primary roots that can penetrate turf or gain a roothold amid weeds. Oak, for example, can colonize grassland. The little seedlings that arise from smaller tree seeds are only able to thrive on ground that is bare or nearly so. Elsewhere they are soon smothered by faster-growing herbs or grasses. This explains why forests, once cleared, return slowly unless re-planted by man. But a natural established forest, left to itself, can renew its kind indefinitely. Whenever an old tree dies and falls, leaving a gap, natural seedlings soon spring up, under the varying conditions of light and shade. They compete for growing space until one takes the lead and replaces the fallen adult tree.

Forests and deciduous trees

To sustain their vigorous seasonal growth, deciduous trees require both fertile soil and ample spring and summer rainfall, combined with adequate warmth and sunshine. These conditions are met in the northern belt of broad-leaved forests that extends across North America, Europe and much of Asia. It lies between the cold arctic tundra and the hot deserts. Because of its fertile soils, this deciduous forest zone has long attracted farmers and settlers who have made clearings in it, to grow their crops or to pasture their sheep or cattle. It has, in consequence, been broken up into fragments by farm land. In the north, and on the upper slopes of high mountains, deciduous forest gives way to coniferous forests of pines or spruces. To the south, it is replaced by evergreen broadleaved forest, and eventually by arid deserts.

Oakwoods

Typical timber trees of the temperate zone deciduous forests are the oaks, which make up the genus *Quercus*. The leading species in Europe is pedunculate oak, *Q. robur*, which bears its acorns on long stalks called peduncles. North America has over 40 species of oak trees. The best-known deciduous one in the east is the white oak, *Q. alba*. All oaks can be recognized in winter by the way their buds are grouped in clusters at the tips of their twigs. This arrangement leads to a characteristic pattern of twigs and branches that spread in all directions and so

Above: Deciduous trees become dormant in winter. Before this occurs the leaves often change to beautiful colours—the result of waste products accumulating—as in this New England woodland.

Left: The yellow sweet chestnut leaves are about to fall but next year's buds are already formed.

Above left: Tree felling with a power saw: two cuts to form a wedge are made on the side towards which the tree is to fall, before making the main cut on the other side of the trunk.

build up a huge spreading crown. This is supported on a stout trunk clad in rugged thick grey bark. Oak leaves have a distinctive lobed outline. Male flowers, which open when the leaves do, in late spring, are catkins composed of groups of greenish-yellow flowers on long stalks. The much smaller bud-like female flowers are wind-pollinated. They ripen, by autumn, as the well-known acorn crop. Each acorn is a single seed or nut, sitting in a neat round woody cup.

Nearly all the acorns that fall, in a good seed year, from a large oak tree are eaten by birds, including pigeons, pheasants, jays and woodpeckers, or by mammals such as squirrels, mice, pigs or deer. Those few that escape destruction sprout next spring as strong deep-rooted seedlings, and oak is well-fitted to colonize forest clearings or abandoned farmland. Many others are actually distributed by the animals that store them for future use, like squirrels and jays, and then forget them.

Every oak trunk or stem of reasonable size holds a naturally durable core of heartwood that is exceptionally hard and strong. Yet it can be readily worked to shape when the right tools are used. From the earliest times man has felled oaks to get timber for building houses, ships and fences, knowing that it would last for scores, or even hundreds, of years, without breakage or decay. Oak forms the uprights and beams of all substantial timber-framed houses, and the roof timbers of cathedrals and barns. It provides the framework of keel, posts and ribs for all large wooden ships built in Europe or America, from the warships of Nelson's day to fishing vessels constructed today. Oak is also a first-rate timber for making attractive strong, if somewhat heavy, furniture; planks are specially sawn to display its beautiful grain.

Because oaks come into leaf late in spring, and cast only a moderate shade, many kinds of smaller plants can flourish in their midst. Oakwoods are the home of bluebells, yellow primroses, pink campions and white windflowers or wood anemones. All these expand their leaves and open their flowers early in the year, when ample sunlight reaches the forest floor, making their photosynthesis effective. Oaks support a remarkably large number of insects that feed on leaves, buds, flowers and acorns. These seldom check growth severely, and rarely kill the tree. These insects in turn provide food for many attractive insect-eating birds, such as warblers, robins, blackbirds, thrushes, nuthatches, treecreepers and jays. Fungi of many kinds grow on decaying oak timber, and insects that attack this substance form the prey of woodpeckers. In short, oakwoods provide habitats for a remarkable range of animals and plants. They make excellent nature reserves, whilst at the same time yielding valuable timber.

Beechwoods

Where the soil becomes poorer, or the climate cooler, oakwoods give way to forests of beech trees. This happens both in Europe where the native beech is *Fagus sylvatica*, and in America where there is a similar beech with somewhat larger leaves, *F. grandifolia*. In Britain, beechwoods are found mainly on thin soils over limestone rocks, on the Cotswold and Chiltern Hills, and the North and South Downs. On the European continent they form the highest broadleaved forests along mountain ranges like the Alps and the Pyrenees.

Beech trees are easily recognized by their smooth grey bark, their slender pointed pale brown winter buds, set singly along the twigs, and their simple oval leaves. Male flowers are grouped in yellow tassels at the tips of short stalks, which arise in little bunches. The green bud-shaped female flowers ripen to fruits that consist of hard green husks, clad in soft blunt spines, and holding shiny brown triangular seeds. Beech seedlings, which begin life with two broad, fleshy seed-leaves, thrive best on bare soil free from weeds. They can grow under considerable shade and are able to colonize oakwoods. As beech trees grow taller, they cast deeper and deeper shade until at length, photosynthesis becomes impossible for any green plant that strives to grow beneath them. The floor of a beechwood therefore becomes a lifeless place, holding no plants and attracting few animals. The ground is strewn with brown faded beech leaves, slowly rotting down to become fertile leaf mould.

Beech timber is pale brown in colour, shot through with little red–brown flecks or plates that represent the rays of the wood. Though it lacks the great strength and durability of oak, it is valued for its hardness, smoothness and the fact that it can be easily worked in any direction. It makes good everyday furniture, particularly chairs and school desks, and is used for mallet heads, plane blocks, short tool handles and other objects needing a handy reliable smooth hard piece of wood. It is a first-rate firewood.

Birchwoods

Everyone recognizes birch by its white bark and slender drooping purplish-brown twigs. Its little leaves are triangular to diamond-shaped, with toothed edges, and hang from thin stalks. Pale green in spring, darker later, they turn to bright gold before they fall in autumn. Male birch catkins, shaped like hanging lamb's tails, become golden in spring, when they scatter yellow pollen. At this time the female catkins resemble little green caterpillars, but by September they have ripened to drooping cylinders, which suddenly break up and release hundreds of tiny winged wind-borne seeds. Seedlings are often found springing up on bare soil; they thrive best on poor sands, where the weeds that might otherwise choke them grow very slowly. Silver birch, *Betula pendula,* is the commonest European kind; the American paper birch is *B. papyrifera.*

Birch trees always demand full sunlight. They cannot tolerate the shade cast by other trees, or even that of other birches. Every birchwood is therefore a light and airy place, with green plants carpeting the ground below. Typical birchwood plants are shrubby heather, heaths, bilberries, mat grass and wavy hair grass. There is usually a sprinkling of delicate flowering plants like yellow cinquefoil and golden cowwheat. Birch forests spread right round the Northern Hemisphere, across Canada, Scandinavia, Russia and Siberia, on the cold poor soils just south of the frozen tundra. They also grow high on mountain ranges like the Alps and the Scottish Highlands. In these regions birch has always been an important firewood and a handy source of timber for small articles like tool handles, broom heads, bobbins and simple furniture. It is also used for veneers in plywood and as pulpwood for paper-making.

Other deciduous trees

Several other trees are found growing, as a rule, in association with oak, beech or birch. Where the ground is particularly favourable, they occasionally form small forests of their own kind. Maples, like the American sugar maple, *Acer saccharum,* are recognized by their lobed leaves, always set in pairs, and their large seeds, also paired, with curved papery wings; their hard wood is used for flooring, furniture and paper pulp. Ash trees, *Fraxinus* species, are easily identified by their large compound leaves, each composed of many leaflets ranked on either side of a long stalk. They have winged seeds that hang down in bunches. Tough ash timber is used for the handles of big tools like spades and axes. Walnuts, *Juglans* species, have similar compound leaves, but are easily known by their fragrance when crushed. They ripen the familiar nut as their seed. Their strong wood, with its beautiful patterns of brown, grey and black, is used for gun stocks, decorative carving and high-grade furniture.

Right: Not all trees are big: the snow willow **Salix nivalis**, *must be among the smallest, with leaves barely 1cm (0.4in) long. Always creeping, it has here found a rock crevice in the Canadian Rockies. Note the tiny reddish 'catkins'.*

HEATHERS AND HEATHLANDS
from moors and mountains

One of the few plant families to have a type of landscape named after it must be the heaths; for 'heaths' can refer to either areas of land dominated by members of the heath family (Ericaceae) or to members of that family. Although horticulturalists might restrict the term heath to members of the genus *Erica* it will be used here to describe the whole family.

Heaths are to be found throughout the world, but more especially in the arctic and temperate regions and in the mountains of the tropics. Very few of the family's 80 genera and 1,500 species are to be found native in Australia.

In size the plants range from small prostrate species such as the creeping azalea (*Loiseleuria procumbens*) and the creeping snowberry (*Chiogenes hispidula*) to *Rhododendron giganteum* which can reach a height of 25m (82ft). But all have simple leaves that are usually evergreen. The few deciduous exceptions include the bilberry (*Vaccinium myrtillus*), the bog whortleberry (*V. uliginosum*), the black bearberry (*Arctostaphylos alpinus*) and some azaleas.

Structure

The flowers appear waxy and are usually regular with the petals joined together below. There are typically twice as many stamens as petals and careful observation will show that these are arranged in two whorls; in the outer whorl the stamens are opposite the petals and in the inner whorl they are opposite the sepals. The anthers of the stamens open by end pores to produce pollen grains which are characteristically joined together in fours.

Many members of this family are pollinated by bees and heather honey from the heather (*Calluna vulgaris*) moorlands of western Europe is justly famous. However, the cross-leaved heath (*Erica tetralix*), which often grows with the *Calluna*, has the doubtful distinction of being pollinated by small insects called thrips which spend their lives inside the *Erica* flower.

The fruit may be a berry or a capsule. Some of the berries are famous for their culinary value; those of the cranberry, bilberry and blueberry are used for sauces, pie-fillings and liqueurs. Some berries, such as those of the temptingly-named strawberry tree (*Arbutus*), are edible but insipid. The North American *Gaultheria procumbens* was at one time the source of oil of wintergreen.

Many heaths have tough waxy evergreen leaves designed to reduce water-loss to a minimum. This is often helped by additional features such as hairs, rolling of the leaves and sunken stomata. It seems strange though that the ericaceous plants with these water-saving characters (xeromorphic) should be the ones that frequently grow on peatlands. Why should plants which grow in waterlogged conditions need to save water? Like a man adrift on an ocean these plants have water all around them but are unable to use it freely. The waterlogged conditions result in an oxygen deficiency in the soil. The roots of the plants cannot function properly and so uptake of water is restricted. Water is, therefore, a commodity not to be wasted.

Peatlands are also very poorly endowed with mineral nutrients, usually relying upon rainfall for all their requirements. With these nutrients in such short supply it is not surprising that ericaceous peatland plants are evergreen, so ridding themselves of the annual necessity of losing their leaves with the accompanying loss of hard-won minerals.

Ericaceous plants have, in common with up to 80 percent of flowering plant species, a second line of defence against nutrient shortage. They have, associated with their roots, fungi, referred to as mycorrhizal fungi. These enter the cells of the root and may supply the plant with nitrogen fixed from the atmosphere, the plant in return supplying the fungus with food substances which it requires.

Although the leaves of many of the xeromorphic heaths, eg heather (*Calluna vulgaris*) and small cranberry (*Vaccinium microcarpum*), are no more than a few millimetres long, some woodland species not needing to conserve water have gone to the other extreme. *Rhododendron sinogrande*, from China and Burma, has leaves over 60cm (24in) long and 30cm (12in) wide.

Habitats

Plants of the heath family grow in the most unpromising habitats. They often dominate European acid peatland vegetation with plants such as *Calluna vulgaris*, *Erica tetralix*, *Vaccinium* species and *Andromeda polifolia*. Mats of *Loiseleuria* and *Arctostaphylos* are often to be found amongst the flowering plants of montane habitats. Some ericaceous plants are even adapted to the extreme conditions of the arctic. *Cassiope hypnoides* is typically found in Iceland and the European tundra in late snow patches.

The majority of ericaceous plants, however, have one great dislike and that is lime in the soil. They are, therefore, said to have a 'calcifuge' habit. In such a diverse family opposites can usually be found and such plants as *Erica sicula*, *E. multiflora* and *E. herbacea* from the Mediter-

Arctostaphylos uva-ursi
Bearberry

Right: A single pine stands in a Perthshire moorland dominated by ling, Calluna vulgaris. *Heather moors occur on sloping ground in areas of moderate rainfall and summer dryness.*

ranean area, *Rhodothamnus chamaecistus* from the Alps and *Arbutus* species of Europe and the Americas are all at the very least tolerant of lime and are, therefore, termed 'calcicole' plants.

There is a saying in Great Britain that 'when gorse is out of bloom, kissing is out of season' for it is unusual not to find some species of *Ulex* in flower. The same could be said for the heaths of a European garden for *Erica arborea* flowers from March to June, *E. tetralix* flowers from June to August, *Calluna vulgaris* flowers during August and September and *E. herbacea* flowers until the following May.

One of the most interesting and best-known ericaceous plants is *Calluna vulgaris*, a single species genus found in Europe and eastern North America. This plant is one predominantly of moorlands and heaths but is also found in open woodlands and on old sand-dunes. The high proportion of broken shells, formed of calcium carbonate (lime), to be found in young dunes may prompt the question, 'How does a calcifuge plant tolerate such limy conditions?' The answer is that the ageing process in sand-dunes washes out the lime leaving neutral soil and it is here that *Calluna* survives. It is also found in acidic peaty pockets on limestone. These develop where humus collects in places watered directly by the rain and not by calcium-charged run-off from the rock.

Man's influence on the heaths

In the wild *Calluna* is a small shrub, on average 60cm (24in) high, with pale purple flowers blooming in August and September. Man has produced hundreds of varieties from this one species with leaves ranging from silvery grey and yellow to the darkest green, single and double flowers in all shades from white to the deepest purple, and mature plants from 10–100cm (4–40in) in height.

From the pollen grains preserved in peat deposits it seems that *Calluna* has benefited from man's deforestation activities which probably began in Europe in Neolithic times, some 5,000 years ago. Through the Bronze and Iron Ages and on into historic times man continued to destroy the native European forests leaving open areas which could develop into heaths. However, during the twentieth century much heathland, particularly in western Europe, has been reclaimed for agriculture.

Calluna-dominated heathland is managed in parts of upland Britain for grouse-shooting. Areas of moorland are burnt on a cyclical basis so that there are always some areas of tall old heather for nesting and short young growth for feeding. This burning of the moorlands has been one of the most important factors in developing the present upland vegetation in

Above: One of the 605 species of Erica which occur in South Africa, out of a grand total of 630 worldwide.

Right: The rhododendrons which are part of the Erica family range from large-flowered Himalayan species like R. arizelum (above) to small ones like the alpine R. ferrugineum (below)

Scotland for, in general, only plants, such as *Calluna*, with underground rhizomes are able to recover easily from the ravages of fire. Such management leads, therefore, to a reduction in the diversity of species.

On the high mountains of central Africa such as Ruwenzori, in company with other plants which have become unnaturally gigantic, are heathers as tall as ordinary European trees.

To the gardener the ericaceous plants provide the flamboyant flowers of rhododendrons, the winter colour of *Erica herbacea* and the ground cover of *Loiseleuria*. To the sportsman they provide the heather of the managed grouse moor. And to all who enjoy the fruits of the countryside they provide a wealth of berries, names such as bilberry, blueberry, cranberry and huckleberry making mouths water around the world.

CRUCIFERS
of cabbages and things

Crucifers possess several common features which make this a more 'natural' family of plants than most. With few exceptions, each crucifer species possesses flowers with four sepals, four petals equally spaced to give a cross (the Latin word *crucis* means a cross), four long stamens and two short ones, and a characteristic seed pod. These features make it easy to determine that a species is a crucifer, but can raise considerable problems in determining which crucifer. The taxonomist has the same problem: estimates of the numbers of crucifer species range from some 2,000 to 4,000, divided into about 200 to 400 genera.

Most crucifers are annual, biennial or perennial herbs, and the majority are restricted to temperate regions of the Northern Hemisphere. Indeed some, such as *Cochlearia* species, are common in the arctic, whence the name given by sailors, scurvy-grass, as they ate it as a source of vitamin C during long ocean voyages. This, perhaps, epitomizes the way in which crucifers have become useful to man. Most are edible and contain a group of chemicals called glucosinolates, which although not restricted to

the crucifers, are another characteristic of the family. When the tissues of crucifers are injured, as they are when eaten, the glucosinolates are broken down by enzymes to a number of volatile chemicals including, notably, the isothiocyanates (mustard oils) and it is principally these which give the crucifers their characteristic pungency. Since ancient times crucifers have therefore been accorded medicinal properties, with or without the justification of scurvy-grass. In addition they have been used as garnishes and condiments. Probably as a consequence of this many have been taken into cultivation, and some have evolved from their wild origins to give our modern crucifer crop plants. These can be roughly divided into four categories: vegetables, animal fodders, oil seed crops and condiments.

Vegetable crucifers
Several vegetable crucifers, such as cabbages and kales (both belonging to the species *Brassica oleracea*) were well-known to the Greeks and Romans several centuries BC. In

Europe, the species was developed further during the Middle Ages to give cauliflower, broccoli (calabrese), kohl rabi and in the early nineteenth century, the Brussels sprout. In Asia, a parallel evolution of *B. campestris* took place to give a similar range of cabbage and kale-like vegetables as well as the more cosmopolitan turnip. Apart from their flavour, which in general has decreased in pungency as the crops evolved, the frost hardiness of *B. oleracea* and the rapid growth characteristics of *B. campestris* have contributed to the success of the vegetable brassicas such that at present 20–40 percent of the areas occupied by vegetables in Eurasian countries contain these species.

Other *Brassica* species are also important as vegetables. A derivative of the hybrid between *B. oleracea* and *B. campestris*, called *B. napus*, became established in northern Europe as the swede and swede-rape in the seventeenth century. The radish, *Raphanus sativus*, like the turnip has been developed in different ways in Asia and Europe. A related species, *B. carinata*—the Abyssinian mustard—is traditionally an important vegetable in Ethiopia, and another, *B. juncea* (brown mustard), is grown in Asia for its leaves. Most of these vegetables have also been introduced by immigrants to Australia and America, where they occupy up to 10 percent of the areas devoted to vegetables.

Use as animal fodders
Many vegetables are also grown as animal fodders. The Romans used the turnip as both animal and human food; and nowadays cabbages, kales, turnips, swedes and radishes are used extensively in both roles. However, the fodder crucifers became pre-eminent in northern European agriculture during the seventeenth, eighteenth and nineteenth centuries when many forms were developed to bridge the winter months between the autumn harvest and spring growth with several benefits. Apart from relieving human hunger, more animals could be kept over the winter, the fertility of the soil was improved and rotation farming could be introduced. At their peak perhaps a million hectares (2.5 million acres) of Britain were occupied by these crops. This has since shown a 90 percent decline, in part due to the introduction of agricultural chemicals, but the continuing place of crucifers in agriculture is assured because of their relatively low production costs and high nutritive value.

Oil seed crops
The crucifers contain large quantities of oils in their seeds, and several species have traditionally been cultivated in those parts of the Old World not possessing other oil crops, such as the olive or the sunflower. The Egyptians used the radish. The northern Europeans used various *Brassica* species, *Sinapis alba* (white mustard), *Eruca* and *Crambe* species (rocket) and *Camelina sativa* (gold of pleasure). The Asians also used *Brassica* species, including *B. juncea*. The oil

Brassica nigra
Black Mustard (left)

Sinapis alba
White Mustard (right)

Above: Watercress, **Nasturtium officinale**, is a crucifer which grows wild in many countries. It is commercially grown in special beds of clean running water.

Right: the wallflower is one of our favourite spring garden plants, which has been greatly developed by skilful selection. The garden wallflower originates in southern Europe, but wild relations—like this Kashmir species— grow in most countries between Madeira and north Africa to the Himalayas.

was used largely for lighting and for culinary purposes, especially in Asia—hence some of the pungency of certain dishes in the Indian sub-continent. More recently, because of the rapid growth and cold resistant properties of the *Brassica* species they are being increasingly cultivated as the only oil seed crop suitable for northern Europe, the USSR and Canada. The oil is used in the preparation of margarine and in certain industrial processes.

Condiments from crucifers

Various condiments are prepared from crucifers. The most notable is mustard which derived its name from the Roman habit of crushing the seed of crucifers in wine must. The burning sensation of mustard when eaten or inhaled is, again, caused by the breakdown of glucosinolates in the seeds. A similar breakdown gives a characteristic pungency to horseradish sauce (made from the crushed root of *Armoracia rusticana*). While the pungency of these condiments is desirable, the same compounds if eaten in excessively large quantities may be toxic. Hence, animals are seldom fed exclusively on rape, kale or any other crucifer material for long periods. Considerable efforts are now being directed towards reducing the quantities of these substances in crucifer seed by breeding new varieties of oil seed rape so that the seed meal, after the oil has been extracted, can be fed to animals.

Other uses of crucifers

In addition to their usefulness as major food and oil crops, a few other uses are found for crucifers. Seakale *(Crambe maritima)* is grown as a salad vegetable. Several types of mustard are used as green manure; they grow rapidly and smother other weeds, and are ploughed in to improve the fertility of the soil. Small industries flourish to supply various green garnishes, notably cress *(Lepidium sativum)* and watercress *(Nasturtium* species). The blue dye woad comes from the dried leaves of *Isatis tinctoria,* which has been cultivated for this purpose since prehistoric times.

Weeds

Some of the characteristics of crucifers which have made them suitable as crop plants such as a rapid growth rate and cold tolerance make them well adapted to be weeds. Thus shepherd's purse *(Capsella bursa-pastoris),* charlock *(Sinapis arvensis)* and wild radish *(Raphanus raphanistrum)* are common and occasionally troublesome weeds of arable land. The problems with the crucifer weeds lie not only with the fact that they compete with cultivated crops, but that they may harbour many of the pests and diseases which may then spread to the crucifer crops. This is another reflection of the similarities between crucifers; virtually all of them share parasites which are almost restricted to their family, including cabbage root fly, cabbage aphid, clubroot (a fungus disease) and certain viruses.

Reproduction

As might be expected with a group of plants with such a uniform flower morphology, the crucifers tend to be similar in their breeding behaviour. Typically, pollen is transferred between plants by insects, and many crucifers possess a biochemical system which ensures that they will not set seed unless pollinated by another plant. Exploitation of this system forms the basis for the commercial production of the F_1 hybrid varieties of Brussels sprouts and cabbages which have become so important in the last decade. Many crucifers possess features which encourage pollination by insects, such as brightly coloured flowers and powerful scents. Hence, several are cultivated for their appearance, such as the candytuft *(Iberis* species), and others including *Arabis* species, alison *(Alyssum* species and *Lobularia* species), stock *(Matthiola incana)* and wallflower *(Cheiranthus cheiri)* are grown for both appearance and scent.

Right: A red cabbage cut through. Cabbages have been eaten since prehistoric times, and the Greeks and Romans cultivated them; the large dense heads are the result of constant selection.

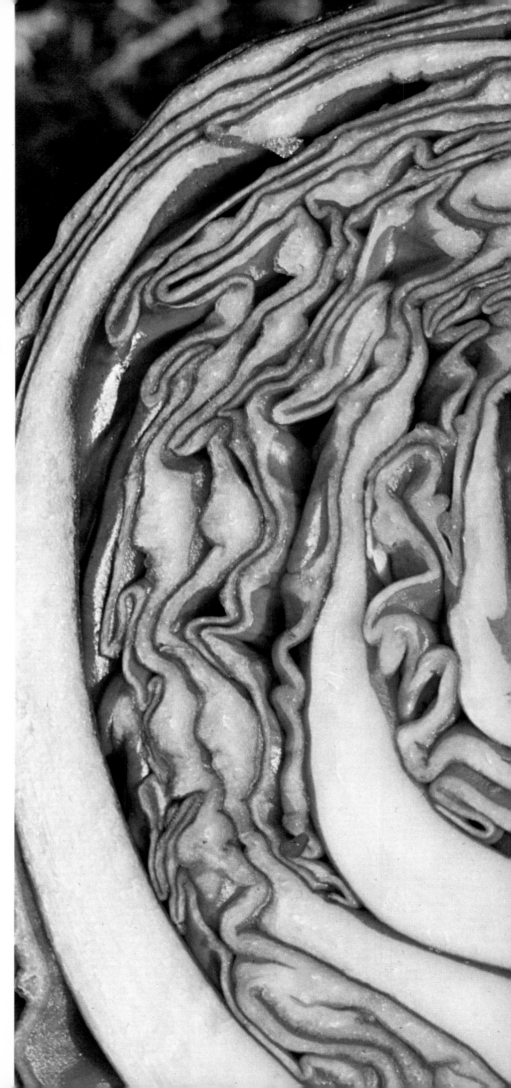

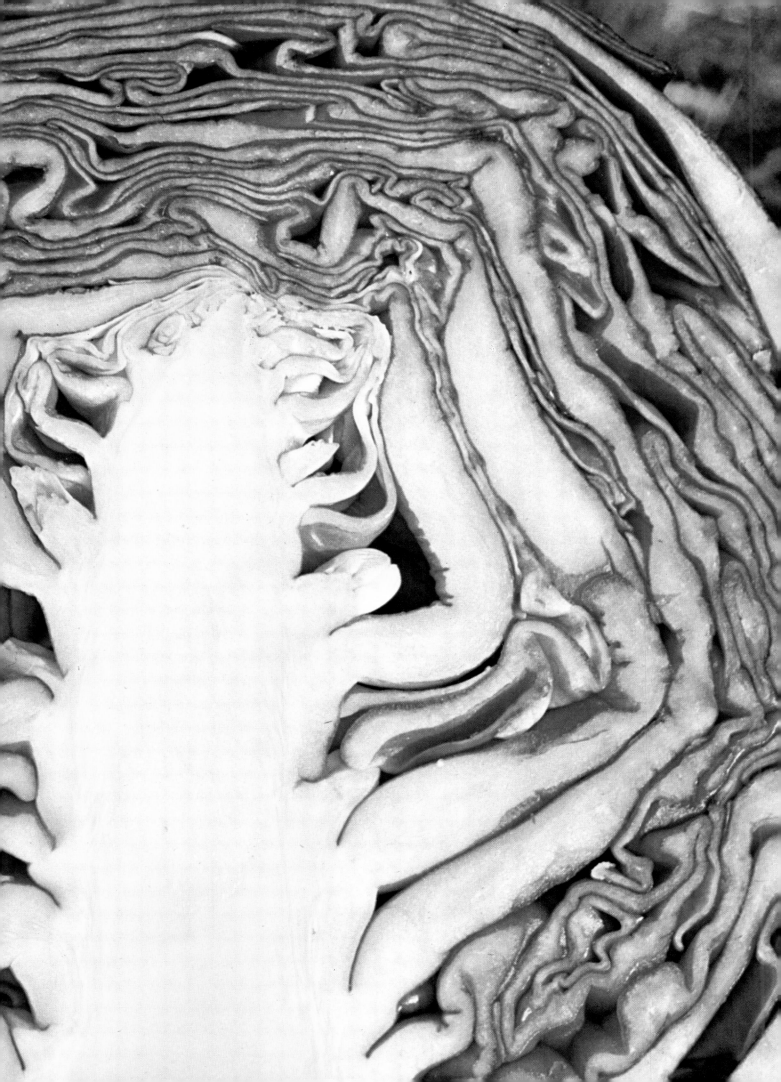

COMPOSITES
multiple flowers mean success

The family Compositae is the largest among dicotyledonous plants. It has a worldwide distribution and members adapted to a wide variety of climatic and ecological situations. The family name is derived from the characteristic way in which the flowers are grouped in 'composite' heads, or capitula, which form a shared base. Although other plant families have small flowers (or florets as they are often called) grouped closely together, the Compositae are unique in that each floret has no calyx as such, although there may be a ring of teeth or bristles. The head, however, is surrounded by several series of bracts, known collectively as the involucre. These initially protect the bud, and during flowering and fruiting may protect the capitulum from insect attack from below.

Flower structure

The five petals of each flower are fused into a corolla tube. The length of the component parts may be equal or obviously different, and some parts may be very small. Flowers with corolla tubes which have five 'teeth' of equal length are called disc flowers, while those flowers which have one or more corolla parts extended well beyond the tube are called ray flowers. The reason for these names can be found by looking at a daisy, which has a 'disc' of flowers in the centre and 'rays' around the outside.

There are three main arrangements of flowers on capitula. The capitulum may be made up of both disc flowers and ray flowers as in daisies, asters, chrysanthemums, marigolds and sunflowers, in which the disc flowers are to the centre with the ray flowers to the outside. The ray flowers generally extend beyond the involucral bracts. Alternatively, capitula may consist entirely of ray flowers as in dandelion, lettuce and chicory, or they may consist entirely of disc flowers and be said to be 'rayless' as are the capitula of thistles, burdocks and cornflowers. Some species, such as buttonweed, Cotula coronopifolia, appear to be rayless but this is because their small ray flowers do not extend beyond the involucral bracts and are therefore inconspicuous. On the other hand, close examination of capitula which appear to have ray flowers around the edge may reveal that they are in fact very large disc flowers. Some species of Centaurea, including knapweeds and cornflowers, are like this. The carline thistles, in the genus Carlina, have large coloured involucral bracts which look like rays.

Reproduction

Within the corolla tube are the reproductive structures. The male androecium consists of five anthers and the female gynoecium of one inferior ovary containing a single ovule. The anthers are fused into a tube and release their pollen on the inside. The style growing outwards from the ovary pushes the pollen out of the tube. At this stage the style has its two receptive stigmatic surfaces pressed close together, giving it a solid rod-like form. Only when the style has grown beyond the anther tube does it 'open' to reveal the two stigmatic surfaces. This very greatly reduces the chances of pollen from an anther being transferred to the stigma of the same flower.

Some flowers may not have any anthers and may be female or sterile if the gynoecium is non-functional. Such flowers are often outer ray flowers. Flowers which are structurally hermaphrodite may produce seed and pollen, pollen alone or nothing at all. They are rarely, if ever, entirely female.

As with the different types of corolla, giving disc and ray flowers, so flowers of different reproductive function may be borne on the same head. In this case the female flowers are to the outside of the head, the hermaphrodite flowers inside them and the male flowers to the centre. The flowers develop and mature from the outside inwards. This means that the outer female flowers are likely to be fertilized by pollen from other heads and from other plants since the pollen from the male flowers on the capitulum will not have been released. If this outbreeding mechanism should fail and they are not pollinated or fertilized, then the pollen released later from the flowers to the centre of the capitulum may be transferred to them. This may be considered as a sort of 'fail-safe' mechanism.

A flower in which the ovule is fertilized by a pollen grain produced by its own anthers tends to produce seed and then offspring very similar to itself because it has only its own genetic material to use. Such plants will survive if they can find an ecological niche similar to the parents. In turn they will produce similar offspring. Although this gives a short-term advantage for the rapid colonization of suitable sites, in the long run it is an evolutionary dead end. Only plants in which pollen from other flowers fertilize the ovules will tend to acquire different combinations of genetic material. Only they will yield offspring which have the potential to survive in different situations. So in the long run an outbreeding mechanism can lead to evolution and survival.

As the flowers on the composite head are ready for fertilization at different times, several

Chamaemelum nobile
Chamomile

Right: The annual sunflower, Helianthus annuus, is widely grown for the oil extracted from its large seeds, used both for cooking and industrially. The American Indians used to grind them into meal.

Above: The outer florets of the knapweed **Centaurea scabiosa** *are in fact extended disc florets, not true ray florets. They are neuter, and their function is to attract insects.*

Left: The huge flower-heads of the globe artichoke, **Cynara scolymus,** *show very clearly the involucral bracts. These are edible, and the plant is widely grown for its heads.*

different pollination events, spread over several days, must take place. This increases the chances that the pollen will come from different plants, which increases the chances that new combinations of genetic material will occur. Thus, a single flowering head may yield seed produced by cross-pollination with several other plants from the surrounding or nearby populations. Such a system is good from the evolutionary point of view as it tends to conserve the genetic information within the population, while effectively 'reshuffling' it.

Dispersal of seeds

The fruit of composites is a hard dry achene. The greatly reduced calyx of each floret often forms a collection of bristles or hairs at the tip of the achene, a structure called a pappus which acts as a tool for dispersal of the achene. Wind catches the pappus and it floats away bearing the achene. In some genera the involucral bracts fold downwards and the achenes, each with its pappus, form the well known 'clock' typical of the dandelion.

Wind dispersal can lead to a widespread ran-

dom distribution of the seeds. This may deposit the seed in new and possibly unfavourable situations. If they have an adequate new combination of genetic material they may be successful and so extend the range of ecological habitats colonized by the plant.

In other genera the pappus is very reduced or absent. Many of these plants have barbs or hooks on the seed, or on the capitulum. If the hooks catch in the fur or feathers of animals the seed may be dispersed to the sites frequented by the animal. If these are fairly constant then the seed may well be distributed to an area similar to its origin. This will aid the rapid colonization of similar ecological habitats.

Adaptations to environments

With these breeding systems for maintaining and producing variation and efficient dispersal mechanisms it is hardly surprising that the family Compositae is so successful. It contains a very large number of species which live in a wide variety of places and have a great breadth of form and structure.

The genus *Senecio* has an African species, *S. adnivalis,* which grows in the Ruwenzori mountains and reaches a height of 6m (20ft). *S. kleinia* from the Canary Islands, has a fleshy stem and grows to a height of up to 1.8m (6ft). In Britain the groundsel species of *Senecio* are generally only several centimetres high and are annual weeds.

The familiar lettuce, is a cultivated variety of a slightly less palatable wild species of *Lactuca.* Many of the composites are attractive to grazing animals. Some genera have reduced their attraction by the possession of unpleasant milky sap

or spiny leaves and capitula as in the thistles. Perhaps the most extreme anti-grazing and anti-dehydration mechanism is that shown by *Launea arboresceus* which can survive attacks from anything bar camels. It is known as camel grass and only produces flowers or foliage leaves within a tangled ball of reduced spiny leaves borne on woody stems. It is the botanical equivalent of barbed wire.

Uses by man

The root systems of composites vary. The dandelion has its well-known tap root, the dahlia has tubers and daisies have fibrous roots. Root types may vary within genera. The genus *Leontodon,* has species with tap roots, fibrous roots and tuberous roots. The possession of root tubers is a great asset to the horticulturist and the gardener, for they enable easy propagation of the plant.

Many composites are familiar as ornamental garden plants. Dahlias, chrysanthemums and asters are widely grown and bred. Greater head size has been but one of the characters favoured by their breeders. In the wild this is often a disadvantage. Not only could the head snap off but greater amounts of seed and genetic material could be lost if insect infestation of the head occurred. However, if a lot of small flowering heads are produced then the damage may be contained to one or a few of them.

Apart from lettuce, chicory, salsify and the ornamentals, the two most important composite genera are perhaps *Helianthus,* the sunflower, the seeds of which yield oil of considerable economic impotance, and *Pyrethrum* which yields the important insecticide of the same name.

The family Compositae is successful and diverse. Much of its success must be due to the capitulum of the flowers which characterize the family. This aggregation of individual flowers into a single head gives the plant an elevated evolutionary status as a 'superplant'.

Tussilago farfara
Coltsfoot

CACTI AND OTHER SUCCULENTS
struggle against drought

Succulents, so-called because of their juiciness (from the Latin *succus* meaning juice), are grouped together on the basis of their peculiar form, evolved in adaptation to the harsh drying conditions of their semi-desert habitat. This is an unnatural grouping as plants of succulent form, retaining stores of water within their swollen stems and leaves, are found in quite unrelated plant families. The Cactaceae, an almost exclusively American family containing many members of typically leafless succulent form, illustrate the wide variety of body shapes and blossoms peculiar to succulents; in other families some resemble stones and others weird sea animals or *objets d'art*. Though most cacti are succulents, there are some genera within the family, such as the genus *Pereskia,* which are non-succulent and have a 'normal' leafy habit. On the other hand, within some principally non-succulent families there is an occasional succulent member, such as the fleshy species of *Senecio* and *Kleinia* in the family Compositae and *Agave* in the narcissus family, showing that succulence is an opportunist mode, evolving in response to a parched environment, rather than a generic one. Other families which have some succulent members include those of the geranium, purslane, vine and milkweed. The two most important families of succulents besides the cacti are the Aizoaceae (mesembryanthemums) and Crassulaceae, both centred on South Africa, although the latter include the hardy stonecrops and houseleeks widely spread in Europe.

Distribution

Succulents are widely but unevenly distributed. Originally most succulent plants other than cacti came from Africa, though some species are derived from southern Asia and some from America; all cacti however are American in origin, their presence elsewhere possibly resulting from the activities of man or birds. The spread of *Rhipsalis* in Africa and Asia is thought by some to have resulted from birds carrying its sticky fruit, while the prickly-pear cactus was brought to Australia by man to serve as range fence and emergency stock feed; before it was brought under control by means of an insect parasite it had taken over more than 24 million hectares (60 million acres). Some *Opuntia* species have penetrated north of the fiftieth parallel, which corresponds to the latitude of Stockholm; in South America some species grow even in the inhospitable climate of Patagonia. In Peru and Bolivia, cacti reach as far as the snow-line, 3,700–4,700m (12,000–15,400ft), such as the mountain *opuntias* which bear woolly blankets of hair to protect them from the cold. Mexico is however the true home of the cactus—the state coat-of-arms bears an eagle holding a snake in its claws, perched on an *Opuntia*. In the valleys north of Mexico City, huge thousand-year-old globose cacti *(Echinocactus ingens)* and the profuse and strange white-haired *Cephalocereus senilis* ('old man of the desert') give the impression of some vast purpose-built garden. Arizona has a characteristic appearance with open forests of candelabra-like saguaro *(Carnegiea gigantea)*, cacti of 15m (49ft) and over in height, real cactus giants. The spherical species average 1m (3·3ft) in diameter and 2m (6·6ft) in height; these can weigh more than a ton! Some of the rarest cacti occur in Mexico and are miniatures of only a few centimetres in diameter; other rare species are still being uncovered in areas explored only recently such as Baja California.

Adaptation to drought conditions

Cacti and other succulents are true xerophytes, adapted as they are to endure drought by having a very slow water loss in transpiration (the rate of transpiration of an *Opuntia*, a typical succulent, was found to be only one-thirtieth of that from an equal area of a thin leaf), the result of a number of specialized adaptations. Within their swollen tissues these plants contain large quantities of mucilages, slimy substances able to retain large amounts of water by imbibition; by acting essentially as little reservoirs of water, these odd plants have become able to obtain their essential nutrients even when there is a water shortage. This enables them to survive long periods of drought even though they appear dead. Cacti of the arid North Chilean desert, lying freely in the sand, can preserve a spark of life even if no rain falls for several years; and a laboratory-kept giant cactus, never watered, was found after six years to have lost only one-third of the moisture it originally contained. Succulents, then, are masters in the economy of water, enabling survival on the very threshold of death.

Evolution of succulents

It is thought that all succulent plants evolved from other related plants growing in a normal environment by their gradual adaptation to

Melocactus communis
Turk's-cap Cactus

Right: Bizarre prickly pears (Opuntia) are very much at home in the extremely arid conditions at low levels on the Galapagos Islands, with volcanic ash to root in.

the changing conditions of their habitat, in particular to the amount and regularity of rainfall, and forms which took to succulence became the forerunners of the present-day extreme forms. The cactus line evolved most probably from ancestors originating in the vast tropical forests in response to the changing climatic conditions of the areas in which they were established, or to which they had spread. In the cactus family, this adaptation to dry environments took the form of losing leaves and developing round or cylindrical bodies. They did this in order to reduce the loss of water by evaporation from the large leaf surface-area.

At the same time the vital processes of assimilation and transpiration, normally enacted by the leaves, were gradually taken over by the green stem, the thin tissues of which came to store as much water as possible for periods of drought. (In many species, the stem tissues are about 90 percent water.) 'Fat finger' shapes were

adopted as the best stem shapes for storing the maximum amount of water with the minimum surface area exposed, to cut down on evaporation.

To further this latter end, the stem skin became much thickened and wax-covered in many species, with the stomatal pores very widely-spaced and depressed in their positions in the skin. Often the cactus skin is ribbed and knobbly as well. This helps the plant to shrink in dry periods; by so doing, it can withstand the loss of as much as one-third of its total weight.

The root system also shows adaptations to desert conditions; many cacti have far-reaching fibrous or tuberous roots, located specially near the soil surface, to take up available water, in the form of dew as well as rain, over as wide a range of soil as possible. Since the rainy season in semi-desert regions usually consists of only a few downpours, the plants need to absorb as much moisture as they can during that period.

In humus-rich soils, which have better water-retaining properties, cactus roots form fine dense networks to exploit this condition. Quite a lot of cacti possess roots rather like turnips, often larger than the stem, which serve as underground water stores. Column-shaped cacti have thick round perpendicular roots in order to provide anchorage in soils which tend to be fine and unstable.

This evolution from 'normal' leafy plants into the forms of cacti so easily identifiable today happened very gradually. Sometimes the various stages of development may be observed in a brief show during the sprout and seedling stages of some cacti. In the very primitive cacti genera, the seedling produces fully-developed leaves like ordinary plants, while in more advanced genera, these appear only as traces in the first few days following germination, later to vanish almost completely. It is likely that the bushy *Pereskia* species which bear normal leaves, growing and flowering

Above: Most cacti, like this small **Hamatocactus** *from Texas and New Mexico, have large showy flowers—essential to attract insect visitors. Here they almost hide the spiny plant.*

Above left: The Saguaro cactus, **Carnegiea gigantea,** *can reach 15m (49ft) tall and weigh ten tons. Such monsters are the trees of the American deserts. In Arizona it grows among free-flowering annuals and drought-adapted shrubs.*

Left: Pebble-like 'living stones' like **Gibbaeum heathii** *show extreme adaptation to very dry conditions in the reduction of the plant to only one or two pairs of tight-packed leaves.*

much like the wild rose, are very close to the original predecessors of the succulent cacti. Today's leafless cacti can no longer produce leaves even if the climatic conditions are changed to provide an ample supply of water, so that when epiphytic cacti occurred in damp forests, their stems merely became flatter and broader until they resembled leaves, thus adapting themselves to the new conditions by increasing the surface area available for evaporation.

However, many non-cactaceous succulents have retained their leaves and here whole series of adaptations can be traced, from 'normal' to extremely fleshy leaves within a family, the most noteworthy example being the Aizoaceae. This family is a living textbook of evolution, the genera exhibiting gradations of form according to the amount of specialization imposed by the habitat. At one end of the scale are more or less woody plants, with well-spaced leaves arranged in opposite pairs; the fleshy leaves are the main indication of adaptation. Then there is a group with trailing stems but leaves more fleshy and more closely packed on the stems. These merge into the really fleshy forms, where stems are almost or entirely absent, with fewer and fewer pairs of leaves. The next stage comes when these leaves are reduced to a single pair, which may be partly joined or, in the most extreme cases, are converted into a single rounded mass or 'plant-body'; the division between the pair is sometimes indicated by a groove or slit, or is sometimes not apparent.

It is remarkable that similar environmental pressures (long hot, dry periods typical of semi-desert or prairie climates) in quite separate parts of the world, caused all the different un-related succulents-in-the-making to evolve along very similar lines, to adopt various expressions of the same strange physique. A good example is the amazing similarity, considering their such separate origins, between American cacti and African euphorbias. There are also similarities between certain cacti and agaves; for instance in some species of the cereus group, which have stems like those of agaves, the similarity is expressed in the name of the plant, as with *Ariocarpus agavoides*. This is a true case of parallel evolution.

Protective spines

Apart from succulence and weird stem shapes, cacti are also noteworthy in appearance for their spines which serve as protection against desert-roaming creatures which might otherwise find a welcome source of food and drink in a juicy cactus. Spines vary greatly. They may be strong spikes of yellow, red, black or brown, 10cm (4in) or more long, or long and silvery hairy outgrowths, as in *Cephalocereus senilis*. Some *Ferocactus* species have large and cruel fish-hooks such as *Ferocactus latispinus,* or the 'devil's tongue', which has thick flat red-covered hooks up to 0·5cm (0·2in) wide. Some *Mammillaria* are covered in delicate white 'feathers', while some *Cereus* species bear a fine 'wool' completely covering the stem to protect against the cold of the night and the heat of the sun, as well as a moisture-retaining net. Cactus spines are really modified leaves and not bristles which are formed from cutin, a substance produced by the plant's skin; bristles are peculiar to the genus *Opuntia,* and they readily break off to cause itching of the skin.

Reproduction

However, the greatest attraction of the cacti is their glorious flowers. These brilliant blossoms, which range through all the colours of the rainbow with the exception of pure blue, can be as large as 40cm (16in) across, among the largest in the entire plant realm, or be as tiny as those of the *Melocacti*. During the day-time these bright flowers attract visiting insects, sometimes even humming birds; when the day-light fails, the huge white nocturnal cactus blooms unfurl and, strangely luminescent in the dark hours, attract the night-flying pollinators, mainly moths and bats, with their strange odours. Flowers burst open at the start of the growing season, emerging from the growing points, generally as solitary blooms rather than clusters. Following pollination, the ovary turns into the fruit which, as with the flowers, can vary considerably through the species. Fruits tend to be more long-lasting ornaments than the flowers and most of them are edible as well.

Some fruits are minute, others reach the size of plums, eggs, even of oranges. In texture they may be fleshy or dry, and upon ripening they may either burst in various manners, or dis-integrate in a characteristic fashion.

Within the fruit pulp lie the cactus seeds, microscopic or hailstone-sized, attached to the walls of the internal cavity of the ovary. Dispersal of the seeds depends on the nature of the fruit. Fleshy fruits are eaten by birds, the seeds passing un-harmed through the gut, whilst ants spread the seeds of ground cacti.

Most cacti can reproduce vegetatively from a number of planes on the body, though the chief centre of this activity is the crown. *Opuntias* can grow roots or shoots from any part of the body, even from tiny fragments of stem, and thus can multiply at an amazing rate. Should conditions be favourable, their stubborn growth can take possession over vast areas in a short space of time; hence the prickly-pear havoc in Australia. Many succulents of other families are equally fast in rooting from pieces of stem or even broken-off leaves.

Use to man

At one time, cacti served many useful and important purposes for man, but nowadays technology has largely overtaken these. In poorer parts, however, the large fruits of some *Cerei,* especially those of *Opuntias,* are regularly eaten by the native Mexicans as 'tunas' and 'cactus figs'. These may be either in a fresh state, preserved with sugar, or fermented into alcoholic drinks. Ripening as they do in the dry season, these fruits can be important when drought has caused other crops to fail. In parts of Brazil, spineless *Opuntias* are used for fodder, whilst in Mexico, some *Cerei,* notably *C. marginatus,* are frequently used to provide thick and impenetrable hedging.

When the giant cacti age, the lower parts and axial tissues turn woody, for these giants, which can weigh several tons, need a firm support. This 'wood', though very light, is tough enough to use for building houses, making household utensils, and for fuel in these otherwise barren and treeless localities.

Interesingly, a significant part in the Mexican economy was formerly played by the production of the red colouring material, cochineal, extracted from an insect bred on some host species of *Opuntias*. Nowadays, however, the cheap aniline dyes have largely replaced cochineal.

Many cacti have important medicinal properties. The massive-flowered *Selenicereus grandi-florus,* or 'queen of the night', for instance, yields a juice containing a glycoside valuable in the treatment of heart diseases. A well-known Mexican cactus, *Lophophora williamsii,* generally known as 'peyotl', contains alkaloids (mescalin, anhalonin, and pelotin) which produce colour and sound hallucinations, and at one time this cactus was the object of a widespread Indian cult.

It is not difficult to see why this strange group of plants, beautiful both in their glorious blooms and in their adaptations to their environment, have become worldwide objects of interest and cultivation.

Right: The South African mesembryanthemums like Drosanthemum speciosum are often incredibly profuse with flowers, which makes them valued as garden plants.

PALMS
the tropical providers

Palms are a large group of trees constructed on a peculiar plan, which makes them look, quite simply, like a huge tuft of leaves set at the top of a long pole. They evolved about 100 million years ago under tropical climates, and with few exceptions, cannot tolerate frost. Nearly all the 3,500 different species are confined to the tropics, hence palms have become symbolic of hot climates.

Structure and life pattern

Botanically, palms belong to the subclass Monocotyledones, that is, they have only one cotyledon, or seed-leaf, within each seed. In the palms, this leaf does not emerge itself from the seed-coat, instead it nourishes a stalk that

emerges and sends down the first root to gain water and mineral salts from the soil. At the upper end of this stalk a bud develops to become the starting point of the palm's upright stem. This seedling stem bears juvenile leaves which are different in form from those of an adult palm. They look rather like broad blades of grass, or the leaves of the wild plantain weed, *Plantago lanceolata*. True leaves come later. All palm leaves develop at the tip of a single stem for, with few exceptions, palms never bear side branches.

Palm leaves are of two main types, according to species. Some, the fan palms, have a broad spreading blade, which may, or may not, divide into radiating arms; it is usually carried

on a long tough stalk. Others, the pinnate or feathery-leaved palms, have a compound leaf with a long central stalk and many side leaflets. Both develop in the same way, from a leaf element that grows at its base, not at the tip as in broadleaved trees. Palm leaves are protected as they emerge by a large leaf-like sheath, which later bends down and falls away. Leaves always expand in strict succession, with the youngest and smallest at the tip, older ones below, in a spiral down the stem. They appear at regular time-intervals of a few weeks. Some reach enormous size, up to 5m (15ft) long on feathery-leaved palms.

Palms are evergreen, and each leaf lives for several years. Eventually it turns yellow, its stalk bends outwards and downwards, and it dies. Certain palms carry a 'skirt' of dead, drooping leaves down their trunks for many years; others drop their dead leaves but keep the bases of the stalks, and others again develop clean stems.

Because growth is only possible from the soft bud at the tip of the stem, any major injury to it means the death of the tree. The stalks of the leaves that surround it are therefore armed with sharp spines, which discourage browsing animals, such as jungle tapirs; some spines point one way, some another. The soft shoot of the coconut palm, which is good to eat, is protected by law in many countries, to ensure the survival of a valuable food source.

The woody stems of palms are built up in quite a different way to those of either broad-leaved or coniferous trees. They consist of fibrous bundles of conductive tissues, set within a hard outer cylinder. Some of these bundles carry upwards the root-sap which consists of water and dissolved minerals. Others carry downwards the leaf-sap, with its sugars or carbohydrate foods, obtained by photosynthesis in the leaves. The two elements run side by side, in spiral paths from leaves to roots. There is no central cylinder of hard wood, no cambium tissue to effect secondary thickening of the woody stem and no annual growth rings. Hence, with a few exceptions, such as the African oil palm, the trunks of palms never get thicker. They remain the same diameter from base to top.

Palm leaves have parallel veins that run from the stalk to the edge. There are no branching veins spreading sideways, such as occur in broadleaved trees. If a leaf is broken into segments by the wind, as often happens, these segments continue living just as before.

There are no large woody roots at the base of palms. Instead, small roots made of bundles of fibres, rather like those seen on flower bulbs, radiate out through the soil from the base of the tree. These roots are immensely strong, and though leaning palms are common, it is rare to see any palm blown over through root failure, even on coasts exposed to hurricanes.

Palm flowers seldom attract notice, for they are usually borne in the central tuft of leaves high up the stem; a few kinds have flower spikes lower down. Some palms bear flowers with organs of both sexes. More usually male

Phoenix dactylifera
Date Palm

Above: Coconut palms (Cocos nucifera), here seen in the Seychelles, are often found along island beaches because the huge seeds can float, which has distributed the plant widely.

flowers open in clusters separate from those of female flowers, but on the same tree; date palms are an exception, being either all-male or all-female. Each inflorescence is protected in bud by a large leafy sheath called a spathe, which falls away later. As pollination is effected by wind there are not bright petals to attract insects, no scent and no nectar. Each inflorescence is branched and carries a large number of separate flowers, each with three green rudimentary sepals and three green rudimentary petals. Male flowers have six stamens each. The female flowers have a pistil likewise divided into three sections, from tip to base. They sometimes bear three, or more, seeds, but usually only one completes its development.

Most palm fruits are single-seeded, so fruit and seed may be described together. Some, like coconuts, are hard nuts that develop in a fibrous outer husk. Others, like dates and oil palm fruits, are plum-like, and consist of a hard stone-like seed, surrounded by sweet or oily pulp. Many are spread by birds or beasts, but some, like the coconut, float and are carried by water, even across oceans. Palm nuts are naturally brown or black, but the soft fruits are more colourful—yellow, red, blue, white or purple.

Under tropical climates many palms have no special seasons for growing, flowering or fruiting. They produce new leaves, open flowers and ripen fruits, at regular intervals of a few weeks, the whole year round. A few exceptional palms, including the sago palm, postpone flowering for several years, then put all their resources into the growing of a huge central flower spike; this exhausts the tree, which then dies. The most spectacular examples—exceptions to the general rule of rather insignificant flowers—are perhaps *Corypha* species, like the talipot palm, which produce an inflorescence 6m (20ft) tall estimated to contain 60 million flowers!

Distribution

In the hot lands around the equator palms play major parts in natural vegetation under widely varying conditions. Though at first sight they look vulnerable to suppression by broadleaved trees, they hold their own in mixed forests; since they do not branch, they grow tall quickly. Some species, like date palms and carnauba wax palms, have low water requirements and thrive in regions with long dry seasons. Others, like the Malaysian sugar palm, tolerate marshy soils, and flourish along river banks even if submerged occasionally by floods. Coconut palms can live on salty sea-coasts.

A fairly well-defined line runs round the globe marking the northern limit of spread for palms. In Europe, it starts in Spain, crosses the south of France near the coastline of the Riviera, and then crosses northern Italy below the Alps on its way east to Asia Minor and the Himalayas; several palms grow in southern China. In America, vigorous outdoor palms are limited to California, Florida and the neighbouring southern states. The hardiest known palm, the chusan or windmill palm, *Trachycarpus fortunei* from the mountains of South China, grows out-of-doors in sheltered places in southern England. It bears fan-shaped leaves, and carries flowers and fruits regularly. Europe has one native palm, a low fan-shaped kind called *Chamaerops humilis*, found on dry hillsides facing the Mediterranean. There is also a form of date palm, *Phoenix theophrasti* (perhaps the ancestor of the cultivated date), native in Crete, which is usually considered part of Europe.

Usefulness to man

In the regions where they thrive, palms prove exceptionally useful to mankind, in fact whole economies are based on them. They provide building materials, fibres, sticks and ropes, and a range of nutritious foods. The peculiar fibrous wood of palms, though sometimes praised in textbooks, is however only used where people cannot get anything better; it is soft and non-durable. The leaves, in contrast, are valuable for constructing and roofing houses. They are cut whilst green, dried in the hot sun and made up into flat panels called, in Malay-speaking lands, *attaps*. These are then used for serviceable windproof walls, and completely rainproof roofing thatch.

The coconut palm, *Cocos nucifera*, is the most widespread cultivated kind. Nobody knows where it first arose, for its seeds have been carried on ocean currents to every tropical sea-shore. The Caribbean zone of Central America is a possible starting point. The familiar hard round brown nut, commonly 15cm (6in) in diameter, ripens within a much larger oval husk, three times that size. This has a grey–brown leathery outer skin, enclosing a mass of tough grey fibre, and the whole provides an excellent float for a big water-borne seed. The fibre, known as coir, is widely used for making matting and cheap ropes. The inner surface of the actual coconut is lined with hard white oily 'flesh', and it also holds sweet white liquid 'milk', refreshing to the taste. Both substances are intended to sustain the seedling when it sprouts on some dry salty beach. The shoot emerges through one of the three dark 'eyes', the other two being 'blind'.

Coconuts are planted everywhere in the tropics beside homesteads, as a handy on-the-spot food supply. There are also big commercial plantations which produce copra, the sun-dried flesh obtained by splitting the nut. This is exported as a source of oil for soaps and margarine. An established tree ripens 50 nuts a year, in a steady succession, uninfluenced by the seasons. The huge feathery leaves, commonly 5m (16.5ft) long, spring from a tall slender stem up to 30m (98.5ft) high, often gracefully curved.

Date palms, *Phoenix dactylifera*, are adapted to the difficult dry climate of the Sahara and Arabian desert fringes, where they are scorched each day but face near-freezing temperatures by night. They need little water, but must have an assured supply from the springs of an oasis, a seasonal stream or a permanently flowing river. Date palms are rather stout trees, rising 20m (65.5ft), with rugged stems studded with bases of fallen leaves. These feathery-compound leaves, grey–green in colour, and often 3m (10ft) long, curve gracefully, and provide welcome shade for desert-dwellers.

Each date palm is either wholly male or wholly female. The Arab grower, unwilling to waste precious water or soil on too many male trees that yield no dates, arranges their sex-life for them. He grows as few males as possible, climbs the trees in the flowering season, takes flowering branches from the males and shakes out their pollen in the crowns of the female trees. The plum-like dates, borne in long-stalked clusters, have sweet nutritious oil-rich flesh and a hard grey central stone—the seed. Easily preserved by drying, they form a handy staple food. Date palms are increased by planting natural off-shoots which preserve the sex of the parent tree; once established, they bear fruit over long spans of years.

The West African oil palm, *Elaeis guineensis*, resembles the date palm but demands a much wetter climate, thriving best under year-round rains near the equator. Its date-like hard-stoned fruits ripen all the year round, and have an oily, rather than sweet, flesh. Oil palms are now grown commercially in many tropical countries for the palm oil that is expressed from both the soft pulp and the actual seed kernel. This is used, like coconut palm oil, for making soap and margarine; it is also a good source of food and cooking oil for local cultivators.

Sago palms, *Metroxylon sagu*, are grown on rich riverside land in Southeast Asia for the starchy food reserves that they build up in their soft trunks, in readiness for flowering and seeding. When, after several years' rapid growth, they are about to flower, the grower cuts them down, splits the trunk and scrapes out the nutritious soft white starch. After washing, this is forced through a sieve and heated gently to make lumps of 'pearl' sago.

The Malaysian sugar palm, *Arenga saccharifera*, holds sap so rich in sugar that its leaf stalks can be tapped, as a source of syrup and sweetmeats. This and many other kinds, including coconut palms, are also tapped to obtain 'toddy', an alcoholic drink. An agile tapper climbs the tree and draws off sweet-sap from a flower spike—which in consequence never flowers. The sugar ferments within a few hours to a weak alcoholic drink; this can be distilled to give a strong spirit called arrack.

Raffia, widely used in basketry, consists of tough fibrous strips pulled from the leaves of a Madagascan palm, *Raphia ruffia*. Piassava, the

Right: Most palms have very numerous small flowers carried in branching spikes. This is the insignificant flower-head of the New Zealand nikau palm, Rhopalostylis sapida.

Below: The palm became a parlour plant when, early in the 19th century, indoor gardening began to be popular. Palms had romantic associations, and scored also because they stood up well to hot, dry room atmosphere.

Above: Coconuts cluster round the crown of the palm. Described as 'one of nature's greatest gifts to man', they supply a wide range of foodstuffs as well as coir from the husks.

Left: The date palm, Phoenix dactylifera, supports millions of people with its very nourishing fruit, as well as having a myriad other uses. Sugar is made from the sap of the wild date.

far tougher fibre used in broom heads, is obtained from the leaf bases of a South American palm, *Attalea funifera.*

Rattans or canes, used both for basketry and the construction of cane furniture, are the stems of climbing palms, *Calamus* species, that grow wild in tropical jungles. The shorter, more rigid Malacca-canes, used as walking sticks, are the stems of a dwarf Malaysian palm, belonging to the same genus.

The Brazilian carnauba palm, *Copernicia cerifera*, which opens large fan-shaped leaves in a climate with a long dry season, is the source of a wax used in polishes of many kinds. The palm develops the wax to restrict water loss. It is scraped from cut leaves and used to repel water in, for example, shoe polishes.

'Vegetable ivory', a hard white strong material used for carving buttons and ornaments, comes from the hard seed of certain palm trees, including the branching doum palm, *Hyphaene theobaica*, of East Africa, the coquilla nut tree, *Attalea funifera*, of Brazil, and the vegetable ivory palm, *Phytelephas macrocarpa*, also South American.

Decorative palms

Many kinds of palms are grown in tropical and subtropical countries as ornamental street or garden trees. In places like the French Riviera, Italy, Florida, California and the Cape Province of South Africa, they give a richly romantic impression of tropical luxuriance. Fairly hardy fast-growing kinds, include the feathery-leaved Canary Island palm, *Phoenix canariensis*, the fan-leaved Californian washingtonia, *Washingtonia filifera*, and the swollen-stemmed feathery-leaved royal palm, *Roystonea elata*, native to Florida.

In the central tropics, where frost never strikes, the decorative kinds include the striking sealing-wax palm, *Crystostachos lakko*, which has brilliant red leaf sheaths, and the slender-stemmed betel nut palm, *Areca catechu*. The latter is the source of the hard nuts that are sliced and chewed, along with a little lime and the leaves of the betel vine, *Piper betle*, by many Malaysians and Chinese as a stimulant, rather like chewing gum.

One of the oddest palms is the double coconut, *Lodoicea maldivica*, native only to the Seychelles Islands of the Indian Ocean. The huge heart-shaped seed weighs up to 25kg (55lb), the largest known for any plant, and takes six years to ripen. The fan-shaped leaves have stalks 10m (33ft) long and blades 5m (16.5ft) long by 2.5m (8ft) broad.

LILIES
neither toiling nor spinning

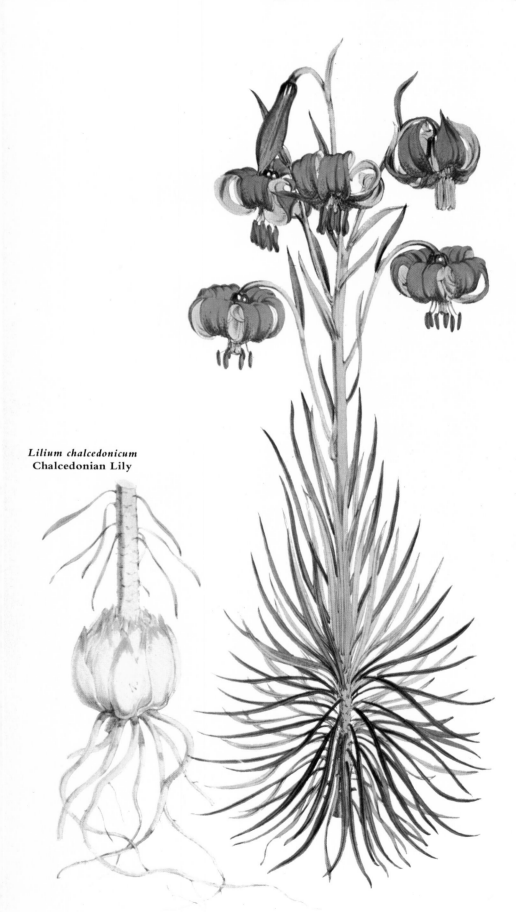

Lilium chalcedonicum
Chalcedonian Lily

The Liliales (lilies) is a large order of plants containing some 8,000 different species in the world, in a dozen or more families. These are widely distributed but are particularly a feature of the temperate and sub-tropical regions.

Adaptations to environments

Characteristically the plants are perennial terrestrial herbs, but there are a few annuals. A number of species in warmer climates are large, woody and tree-like in their growth. There are a few climbers and one family, the Ponte-deriaceae, are aquatic herbs that include the notorious cosmopolitan weed water hyacinth, *Eichhornia crassipes,* which bears colourful spikes of lilac flowers above the rafts of floating leaves.

A striking feature of the group is their variety of adaptations to survive inhospitable climates or seasons. Several families have evolved bulbs, corms or tubers as underground food stores which enable the plants to survive unfavourable conditions. For example the colourful bulbous and tuberous plants such as tulips, crocuses, irises, fritillaries, garlics and grape hyacinths are all characteristic plants of the hot arid Mediterranean garigue. These plants largely contribute to the colourful spectacle of this landscape in the spring, which quickly gives way to scorched stony hillsides during the summer months.

The agaves of America and aloes of Africa are able to survive in dry desert climates by means of their tough succulent leaves and stems which store water and are resistant to water loss.

Flower structure

Flower structure in the order is usually fairly uniform with a corolla composed of six perianth segments. The three outer segments are the sepals while the three inner segments are the petals, but normally they are closely similar in shape and colour. The flowers may be large, showy and solitary as in the tulips, *Tulipa* sp., or they may be borne in long racemes of small flowers as in the Mediterranean plant sea squill, *Urginea maritima.* The latter is remarkable for its enormous bulbs which may be as much as 15cm (6in) across and may project out of the ground on dry stony hillsides. The flowers of the onion and garlics, *Allium* spp., are relatively small but may be clustered into large globose heads. The European wild crow garlic, *Allium vineale,* does not often produce normal flowers but has a small spherical head of tiny bulbs or bulbils which sprout young leaves before they are shed to grow into adult plants.

Pollination

Lily flowers are usually pollinated by insects. The perianth segments normally form a tubular-

Right: The giant yuccas of the American deserts are members of the lily family, as examination of individual blooms will show. They have a highly specialized pollination adaptation.

Above: The delicate 'Mariposa lilies' like **Calochortus splendens** *also belong to the lily family, and—as with most of this very attractive family—many species are cultivated.*

Left: The orange lily, **Lilium bulbiferum,** *in a lush alpine meadow. Some varieties grow on cliffs. Specially large and brilliant forms of this flower have long been cultivated in gardens.*

shaped flower at the base of which nectar is produced. Visiting insects have to push past the stamens to reach the bottom of the tube. In the temperate genus *Fritillaria* the flower is a hanging bell formed from six free perianth segments which produce nectar in grooves at their bases. The stigma, which projects beyond the stamens, matures first so that the chances of cross-pollination by visiting bumblebees are increased.

In the nodding star of Bethlehem, *Ornithogalum nutans,* the filaments of the stamens form a tube while the perianth segments are spreading for display.

Several members of the order have unusual pollination mechanisms. In the American yuccas of the family Agavaceae there is a remarkable relationship between the flowers and their pollinator, a small tineid moth. The creamy-white flowers of the yucca are produced in showy inflorescences and are sweetly scented at night to attract the moths. The female moth visits the flowers descending into one and climbing the stamens one by one to collect pollen with her maxillary palps. The moth then flies to another flower where she inspects

the ovary and if the flower is suitable she bores into the ovary to lay an egg. She then climbs up to the stigmas which form a tube and thrusts pollen into the tubular structure. Unpollinated flowers die quickly but pollinated flowers develop to ripe fruit, part of which nourishes the larval moths. The plant therefore provides food and shelter for the moth which in return pollinates the flowers.

Use to man

A few members of the order produce useful or economically valuable products. The extraordinary tree-like dragon tree, *Dracaena draco,* which is a native of the Canary Islands, and another member of the same genus produce red resins known as dragon's blood which are used in varnishes.

The meadow saffron, *Colchicum autumnale,* of Europe and North Africa yields an alkaloid drug colchicine which is extracted from the corms and is used medically for treating rheumatism and gout. The drug is better known for its controlling effect on chromosomes which has facilitated the study of this aspect of cell science.

Several members of the genus *Allium* are important as food crops or flavouring materials. The onion, *A. cepa,* which probably originated in Asia, has been an important and widely cultivated *Allium.* The bulbs are eaten raw or cooked, and are widely appreciated for their characteristic pungent flavour. Chives, *A. schoenoprasum,* widely distributed in the north temperate region, has tufts of longitudinal leaves and heads of pinkish purple flowers. The mild flavoured leaves are used principally for garnishing. The closely related leek, *A. ampelo-*

prasum var. *porrum,* originated in the eastern Mediterranean and is cultivated as a vegetable.

A number of yams of the genus *Dioscorea* are of local economic importance in the tropics. These predominantly tropical herbaceous climbers such as *D. alata* produce edible starchy tubers which are a staple food in parts of West Africa.

Several species of agave, natives of the arid and semi-arid tropics of Central America, especially Mexico, are cultivated or harvested for their fibres. These perennial plants produce massive flower spikes from basal rosettes of stiff spiky leaves which yield long hard fibres. Sisal, *A. sisalana,* produces coarse fibres which are particularly useful in the manufacture of twines and cordage, while the chopped fibres may be used as a re-inforcing material for plaster-boards.

The related century plant *A. americana* is often planted for its decorative foliage. The majority of garden plants belonging to the lilies are small herbs that are grown for their colourful and decorative flowers. These include all the spring bulbs and related flowers that are such a feature of gardens in north temperate countries. Very early flowering species such as the snowdrop, *Galanthus nivalis,* come into flower during the winter months. These are followed in early spring by the first flush of colour provided by a variety of cultivated crocuses which originate from southern Europe. The daffodils and narcissi, belonging to the temperate genus *Narcissus,* also flower at about the same time. Tulips, which are late spring bulbs, were introduced from western Asia to Holland and other west European countries in the sixteenth century, where they have been cultivated ever since. Tulips are grown commercially for their flowers in Holland, the Fenlands of eastern England and in America. A variety of true lilies, genus *Lilium,* which are cultivated as garden flowers, originated in temperate Asia, especially China, and Japan, and also in America. The late flowering bulbs of European gardens are mainly South African in origin such as the spectacular blue flowered *Agapanthus africana* and the rose pink nerines.

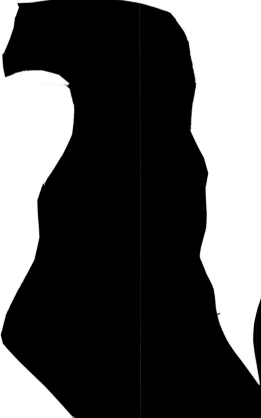

ORCHIDS
beautiful and bizarre

Orchids are fascinating plants. They depend on a peculiar association with fungi during their young stages; they may take years to reach maturity; they may be exceedingly localized or rare; and above all they have some of the most beautiful, colourful and incredibly bizarre flowers known to man. Some are extremely difficult to cultivate. These characteristics perhaps explain the interest and the excitement that is conjured up by the word 'orchid'.

As a group the orchid family is one of the largest families of flowering plants in the world with up to 30,000 or more species belonging to nearly 800 genera. Orchids grow all over the world but the majority of species occur in the warm humid tropics and their numbers dwindle towards the poles. Nevertheless a few species occur within the Arctic Circle in Alaska, Greenland and Siberia, while other species occur above the tree line on mountain ranges.

Structure

All orchids are perennial herbs, but they exhibit an astonishing variety of size and form. The most characteristic features are those of flower structure. The flower is composed of six perianth segments in two whorls. The outer whorl represent the sepals, while the inner whorl represent the petals. One petal (the median posterior) is larger and distinctly different from the others and forms the lip (labellum). As a result of the twisting of the ovary through 180° this lip usually points downwards in the centre of the flower. In most orchids the reproductive organs are borne on a single structure called the column, which is formed by the fusion of the stamens and style. The complex structures of the orchid flower are related to the mechanism of pollination.

The orchid fruit is a capsule formed by the fusion of three carpels. When the fruit is ripe longitudinal slits appear, opening from the tip, to release innumerable minute seeds. These seeds are remarkable for their extremely small size (three million seeds weigh only one gram) and their lack of both food reserve (endosperm) and a well-formed embryo plant. An almost incredible four million seeds are produced by a single fruit capsule of the orchids *Cynoches* and *Anguloa*.

Habitats and adaptations

The largest number of orchids are epiphytes growing on trees in tropical rainforests. They are most abundant in the extremely humid cloud forests of mountain slopes which create ideal conditions for epiphytes with the trees densely clothed with mosses, an abundance of water and an angle of inclination of the ground which allows a good penetration by sunlight. Temperate species and a few tropical ones are terrestrial herbs that inhabit grasslands, woodlands, marshes and bogs, or rocky mountain ledges. A few woodland species of temperate zones, such as the coral-root orchids, *Corallorhiza,* are saprophytes which lack the green pigment chlorophyll and which cannot produce their own food. As a consequence these plants are dependent on the decaying humus in the soil.

The epiphytic orchids of the tropics grow on the branches of trees for support. The relationship is not a parasitic one for the orchid derives no food from the tree except for substances in the decaying outer bark. These orchids have specialized roots that adapt them for living in such an inhospitable place. Clinging roots attach the plant to its support. The network of roots that forms between the plant and the branch entangles dead leaves and other debris and retains a reservoir of humus into which the absorbing roots grow. There are also aerial roots that hang down in festoons. These roots contain green tissue and serve to assimilate. The fleshy living tissue is covered with spongy dead tissue that absorbs water during periods of rain or heavy dew but which protects the living tissue from dessication during dry conditions. All epiphytes have special arrangements for shedding their leaves with a joint where the leaf blade meets the sheath. In parts of the tropics that experience a pronounced dry season these epiphytes may lose their leaves and hibernate as pseudobulbs. A pseudobulb is formed by the thickened stem internodes which are usually formed each year and which store water and food reserves. Some epiphytes store water in their fleshy leaves. A West Indian orchid, *Dendrophylax,* has no leaves at all and is reduced to a series of broad fleshy green roots. Similarly the Malaysian *Taeniophyllum* has a very short stem (about 3cm — 1.2in) with flat green spreading roots on the bark of a tree, and with leaves reduced to tiny brown scales.

Growth

Most orchids, whether they are terrestrial or epiphytic, have the same growth form. Each leafy stem, or pseudobulb, is limited in its growth. The rhizome may be long or short and the pseudobulbs may consist of one or many internodes. Flowerstalks are produced by lateral buds which may arise at the top of the pseudobulb, on its side or at its base. This type of growth with a succession of stems of limited growth is called sympodial growth.

The alternative growth form, termed monopodial, occurs in the *Vanda* tribe of Southeast Asia and in other groups. For example, in the Malaysian scorpion orchids, *Arachnis,* the tip of every stem goes on growing indefinitely, producing roots along its length at intervals, so that

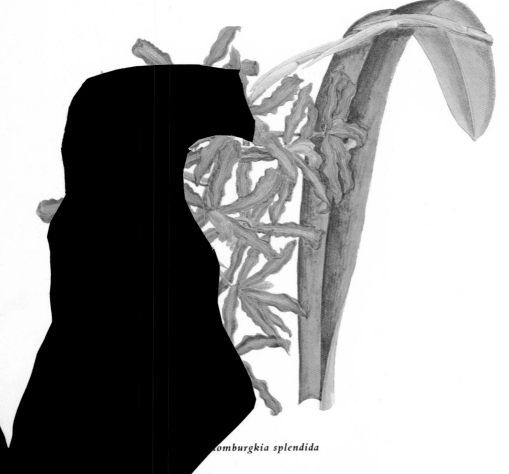

omburgkia splendida

*Right: This **Oncidium** clearly shows the typical formation of an orchid flower, with two horizontal petals and a large yellowish lip, the remaining segments being the sepals.*

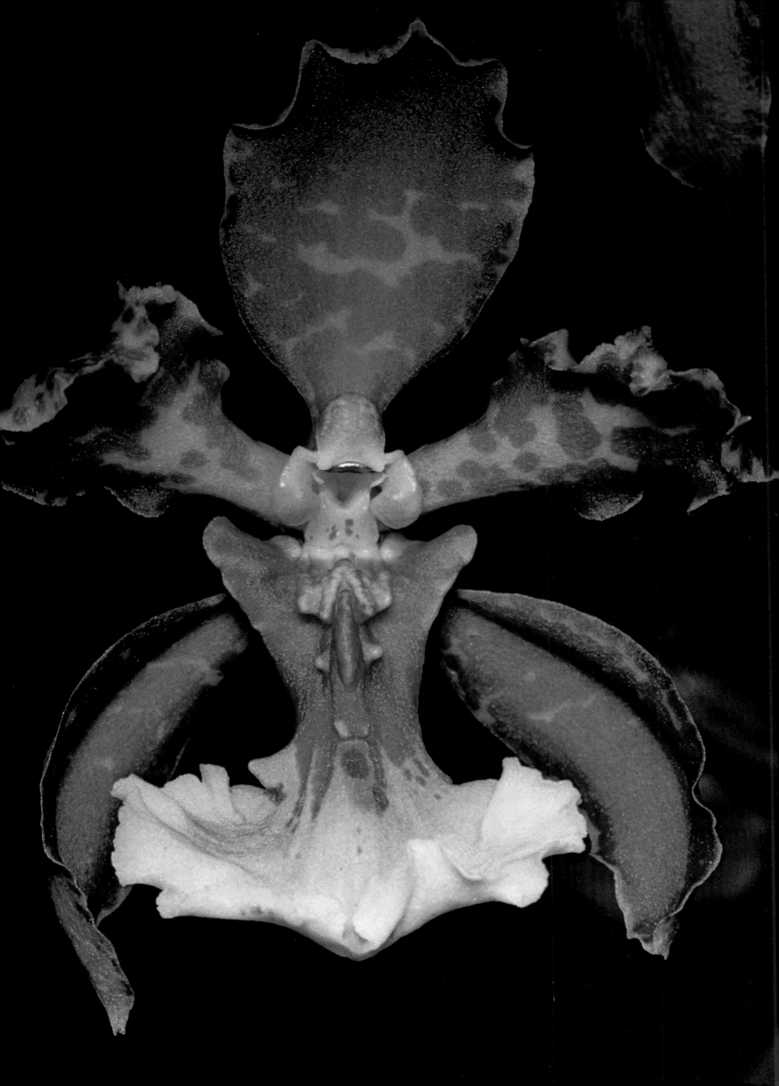

the older parts may die while the tip continues to grow. Some of these orchids develop as climbers or vines if their stems are very long; for example the vanilla orchid, which is grown commercially, forms a vine-like plant.

The world's largest orchid plant is the giant Malaysian Tiger Orchid *Grammatophyllum speciosum* in which the pseudobulbs are 2m (6.5ft) in length with leaves all along, but which occasionally reach twice that size. The individual flowers are up to 12cm (4.7in) across and are borne in huge inflorescences of numerous flowers. These plants frequently grow in the crowns of large trees with their pseudobulbs arching down in graceful curves. The stiff branched roots grow up and outwards to accumulate debris and dead leaves.

Terrestrial plants may only be a few centimetres high, such as the north temperate bog orchid *Hammarbya paludosa*, but other species may be very bizarre in form. The Malaysian bamboo orchid, *Arundina*, closely resembles a clump of bamboo with the flowers borne on slender solid stems.

Saprophytic orchids

A number of unrelated orchids inhabiting damp woodland situations have evolved as saprophytes. These plants have a yellow or pinkish colouring and have very branched fleshy rhizomes which absorb food material from the decaying humus in which they live. The form of the root gives the name 'coral root' to some of these orchids. The evolution of some species as saprophytes is not very surprising when it is realized that all orchids start their lives as saprophytes dependent on a strange relationship with a fungus called a mycorrhiza.

While the microscopic orchid seeds admirably serve the function of wide distribution by wind dispersal, even to inaccessible locations high on forest trees, they sacrifice the food reserve which most plants have to start their development. The orchid seed is dependent on the presence of a fungus growing within its tissues for its germination and subsequent development. The mycorrhizal fungus is a saprophyte which invades the orchid seed and penetrates its tissues. The fungal hyphae are in contact with both the plant tissues and the surrounding soil and provide the orchid seedling with essential nutrients. The relationship between the orchid and the fungus continues throughout the life of the plant but is most essential to the orchid in its young stages before it can produce its own food. Mycorrhizal relationships are widespread in nature and occur also with forest trees and many other woodland plants. The growth and development of the orchid seedling or mycorrhizome is very slow even with the assistance of the fungus. In temperate regions green leaves may not be produced for two or three years and several years may elapse before the plants produce flowers. In the European burnt tip orchid, *Orchis ustulata*, it may take as long as 14 years from germination of the seed to the production of a flower spike.

Pollination

The flower is the most remarkable feature of an orchid. The flower size may vary considerably from the tiny flowers of the epiphytic orchids *Pleurothallis,* from the New World tropics, which are barely 2mm (0.08in) in diameter to the flowers of a tropical American *Brassia* which may exceed 38cm (15in) in diameter.

The fantastic shapes, colours, textures and scents of orchid flowers are all designed with their respective pollinators in mind. Bees and wasps are the most important pollinators of orchid flowers, flies and other insects follow, with birds such as humming birds pollinating relatively few species. Each type of flower is cleverly designed to attract a certain type of creature to visit the flower. To avoid hybridization many orchids are designed to attract only one species. The colour and/or scent of the flower attracts the pollinator, while the complicated flower structure guides and manipulates the pollinator so that the pollination mechanism may work with precision.

The fascinating and intricate processes of pollination in orchids were first investigated and understood by Charles Darwin who published a book on the subject in 1862 (*The various contrivances by which orchids are fertilized by insects*). In orchid flowers the pollen grains are generally in the form of a pollen mass or pollinium. This pollen mass is detached by a visiting pollinator and is transferred to the stigmatic surface of another flower. The pollen mass ensures that when the fruit is formed it will contain a large number of fertile seeds.

Pollination mechanisms

A simple example of the pollination mechanism is given by the European twayblade, *Listera ovata.* The greenish flowers attract ichneumons and other insects that alight on the strap shaped lip (labellum). The insect crawls up the lip to feed on the nectar secreted by a groove

Above: A great many orchids of tropical rain forests are epiphytic, clinging to tree branches with aerial roots which absorb moisture from the air. Many are small flowered.

Above left: Ophrys speculum is one of the Mediterranean bee orchid tribe which mimic insects in appearance, odour and feel, so that male insects are deluded into thinking they are females.

Below left: Orchids of temperate climates are mainly terrestrial, like this Orchis species growing in an alpine meadow in Kashmir at nearly 3,000m (10,000ft) altitude.

running up the centre of the lip. The pollinia are free in the open flower and are supported by a part of the column called the rostellum. When the insect touches the rostellum it explosively produces a drop of sticky liquid which contacts the tips of the pollinia and the insect and sets hard in a few seconds. The pollinia are then cemented to the head of the insect. After removal of the pollinia the flower alters by the straightening of the rostellum so that the stigmatic surfaces are exposed to the next insect visitor. If it bears pollinia from another flower it will deposit them on the stigma.

Orchid flowers generally attract insects by the presence of nectar, but nearly one-third of the family have nectarless flowers and use other means of attracting insects. One of the most unusual types of attractant is found in the waxy flowers of the tropical American orchid Coryanthes that exploit the sexual response of certain bees. The lip of the flower is in the form of a bucket which accumulates liquid secreted by the column. The flowers attract male euglossine bees of the genus *Eulaema* by the production of scent which the bees collect with the tarsal brushes of their front feet. In the process of collecting this scent from the base of the lip the bees become intoxicated and fall into the bucket of liquid. The only escape is via a narrow passage in which the bees pick up the pollinia. After removal of the pollinia subsequent bees can escape much more easily. Those carrying pollinia pollinate the flower in the process. This mechanism appears to be designed to control the movements and reduce the activity of the bees so that they can be guided by the flower and the pollinia can be applied and removed with a high degree of accuracy.

To increase the chances of cross-fertilization the epiphytic Malaysian pigeon orchid *Dendrobium crumenatum* produces spectacular displays of simultaneous flowering. Research has shown that the flower buds developing in the axils of the leaf sheaths stop growing at a certain stage and await a suitable stimulus. Flowering is triggered by a storm which causes a sudden drop in the day temperature. The flower buds then resume development and nine days later the fragrant flowers open just before dawn. Bees visit the flowers in the morning and emerge with pollinia attached to their heads to deposit them on the stigmas of the next flower. The flowers close up by the afternoon and are over by the next morning. The synchronization of the flowering is important for a flower which is so short-lived while the rain, which provides

the necessary stimulus of falling temperature, also provides an ample supply of water for flower formation and subsequent evaporation loss from the flowers.

The European orchids of the genus *Ophrys* exhibit some of the most remarkable pollination mechanisms found in any type of flower. The mirror orchid, *Ophrys speculum*, from the western Mediterranean provides a good example. The lip of the orchid flower is shining blue in colour with a yellow border fringed with red hairs. The flowers superficially resemble female wasps of *Campsoscolia ciliata*. The males of this species emerge some while before the females and are then attracted to the flowers by their scent and appearance. Male wasps alight on the flowers with the head below the rostellum and make repeated and vigorous attempts to copulate with the flowers. This activity nearly always dislodges the pollinia, which become attached to the insect's head. Other *Ophrys* species attract the males of other insects in the same way and sometimes the scent of the orchid flower may be more attractive to the male insects than their own females.

In general orchids pollinated by bees have fragrant flowers with bright colours; those that attract moths are scented and light coloured with long tubular nectaries; butterfly-pollinated flowers are upright with fragrant colourful flowers; while bird-pollinated flowers are usually scentless. Orchids that attract flies often have dull brown or purple coloured flowers and produce odours that resemble decaying flesh.

Flowers that fail to attract any pollinators

may resort to self-pollination. A mechanism in the flower ensures that the pollinia fall down into contact with the stigmatic surfaces, eg in the European bee orchid, *Ophrys apifera*. Species with small insignificant flowers such as the small white orchid, *Leucorchis albida*, are regularly self-pollinated in the bud with the pollen masses breaking up inside the flowers before they open.

Economic importance to man

Cultivated varieties of orchids may produce very spectacular sprays of beautifully coloured flowers. About 50 different genera are commonly cultivated such as the tropical American *Cattleya* species and the *Cymbidium, Vanda* and *Dendrobium* species of Southeast Asia.

The only species of economic importance is the Mexican vanilla orchid, *Vanilla planifrons*. This vine-like plant is grown in tropical countries which have a regular dry season such as Madagascar, the Seychelles and Reunion. The pods are picked and carefully cured by a process of slow drying and fermentation to bring out the full flavour which is imparted by crystals of vanillin that form in the tissues of the pod. The vanilla 'beans' are used intact for flavouring or for making vanilla essence.

Below: An extraordinary colour combination in the Malaysian Coelogyne mayeriana, which is pure green with black markings. Insects are attracted by strong fragrance.

Creating unnatural liaisons

Wild orchids began to be collected extensively for cultivation during the last century, when greenhouses began to be efficient. It was not long before growers were breeding from the new introductions, and producing hybrids between species of very different origins which could not possibly occur in nature. In some cases hybrids between several different genera were created, like the trigeneric *Brassolaeliocattleya* shown here. Though first of all very much a rich man's hobby, orchid growing is now popular among all groups, with as much interest in small-flowered species as in highly bred "bosom orchids" like that pictured here.

GRASSES
the ubiquitous plants

Of all the families of flowering plants, the Gramineae, the grass family, must rank above all others, exceeding as it does in amount, variety and value of its products, and in the number of its individuals. There are about 620 genera and 10,000 species of grasses, varying in size from the lilliputian to the gigantic, and their products form the staple diet for mankind in many parts of the world; the genera *Triticum* (wheat), *Avena* (oats), *Hordeum* (barley), *Secale* (rye), *Zea* (maize) and *Oryza* (rice) have been cultivated for thousands of years.

Distribution

Grasses are distributed through almost every possible kind of environment. There are species thriving in and around water, others which grow in deserts, and yet other species can be found surviving the intense cold of the polar regions. Great adaptability is shown even within a species; the sheep's fescue *Festuca ovina* occurs both at sea-level in Britain and at 5,500m in the Himalayas, whilst sweet vernal grass, *Anthoxanthum odoratum* flourishes on sand, loam or clay

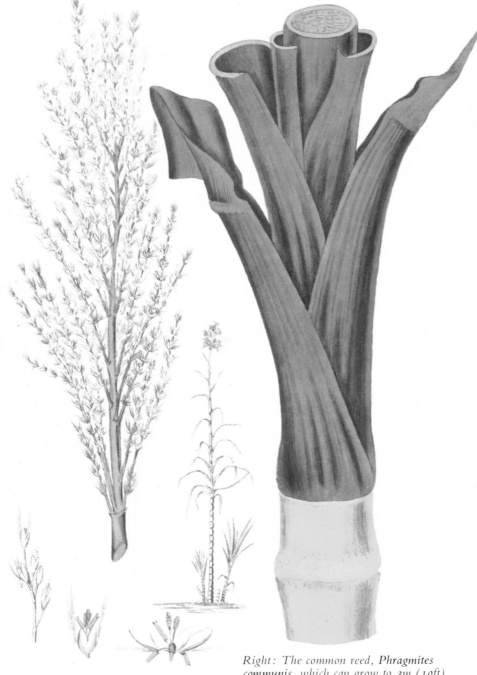

Saccharum officinarum
Sugar Cane

Right: The common reed, Phragmites communis, which can grow to 3m (10ft), seen stacked for thatching; it was at one time extensively used for roofs in some areas of Britain.

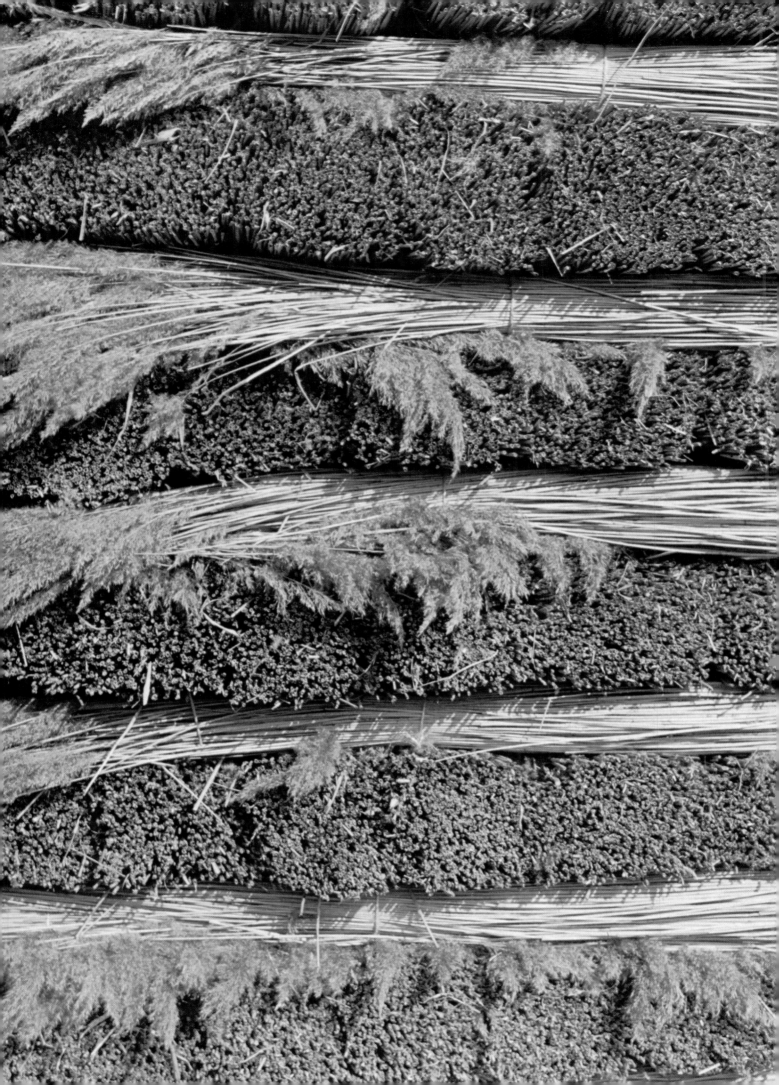

Marram grass and the sand dune system

A typical seashore sand dune system, itself often based on shingle bars, can reach 12–17m (40–50ft) in height in Britain, and can advance remorselessly under the influence of wind to cover whole villages and alter the course of river mouths in the space of a century or two. Dunes are eventually halted when stabilised by plant colonisation, as seen at Studland, Dorset (above). The first major coloniser is often sand couch grass, *Agropyron junceiforme*, which tolerates occasional submergence in sea water. This is typically followed by marram grass, ***Ammophila arenaria*** *(seen, right, with ling,* ***Calluna vulgaris***), before it can support less tenacious plants. At Studland the beach and dunes are backed by heath, where plants like the ling and various sedges and rushes gradually supersede the marram grass. Eventually the heath gives way to shrubs. Many flowering plants exist within these frameworks. Marram grass thrives in fresh sand and moving dunes, its powerful horizontal shoots anchored by roots which can penetrate 18m (60ft) down. Once dunes become static, marram grass tends to disappear.

Above: Sheep grazing on New Zealand grassland remind us of the Biblical saying 'all flesh is as grass', and that every animal life cycle depends initially upon plant life of some kind.

soils, and can be found in countries as different in their climates as North Africa and Siberia. Strains resistant to certain toxic metal wastes have recently appeared with the result that grasses have even become established on mine spoil heaps.

Natural grassland, the home of the fastest animal runners in the world, occurs on all continents, principally in their interiors, as in the prairies of North America, the Argentine pampas, the steppes of the Old World and the South African veld. Here grass species have come to dominate the flora over vast stretches of land. This is mainly because of the climate: rainfall is insufficient to support trees but high enough to keep deserts from forming. In these regions, tough coarse grasses of the genera *Sporobolus*, *Stipa*, *Elymus* and *Agropyron* in particular grow in thick profusion. Their matted roots and rhizomes form a thick turf, often a quarter of a metre or more in depth, which traps and conserves the scant moisture well. These grasses may grow as tall as 3m (10ft) in 500–625mm (20–25in) of annual rain.

Grassland and evolution of grazing mammals

Once natural grassland covered 42 percent of the earth's surface, providing a habitat for the evolution and support of large herds of grazing mammals and colonies of rodents like the prairie-dog. Early hoofed mammals similar to our present-day grazing animals existed some 70 million years ago, but teeth studies show that they were largely browsing animals, cropping forest leaves. Fossil evidence indicates that around some 40 million years ago, at the opening of the Miocene period, a drastic change in the climate occurred. Rainfall decreased, so that forest cover could no longer be supported, and consequently grasses along with other low-growing plants became dominant over vast plains, both in the Old and the New World. In this new habitat, the true grazing animals evolved. Antelopes and sheep are recognizable from the Upper Miocene, and oxen, goats and horses from the following Pliocene period.

Nowadays much of this natural grassland has disappeared. Unwise cultivation practices and excessive grazing have caused this in part, for once the soil's protective covering of leaves and stems and the binding network of roots and rhizomes are destroyed, wind and rain soon erode the soil and dust bowls result. However, in tropical and other regions normally capable of supporting taller vegetation, grassland exists chiefly through man's activities, be these burning or grazing practices. Should these activities stop, the vegetation would revert to its natural tree-covered state.

Structure and growth pattern

Grassland today is principally used for grazing domestic animals and for providing cereal crops for mankind, but to see how grass can be of such value to man, we must look at its form and growth patterns.

Most grasses are annual or perennial herbs with occasional woody forms. They generally have fibrous root systems and can reproduce themselves vegetatively by means of underground stems as in the couch grass *(Agropyron repens)* or by runners like the common bent *(Agrostis tenuis)*, for example. Grass leaves are characteristically linear and are arranged in two ranks on the stem, each leaf having a sheathing base which surrounds the stem. At the top of the sheath, there is a little membrane called a ligule where the leaf blade starts.

Grass flowers are usually hermaphrodite and are carried in groups called spikelets, with enclosing bracts called glumes. Each flower has two protective bracts called the palea and the lemma, with sometimes a long bristly awn attached, as in oats, barley and bearded wheat. Being wind-pollinated, grass flowers tend to be very inconspicuous, lacking petals and sepals (though in the majority of cereals self-pollination occurs). The stamens, usually three in number, are long and hang out of the flower, exposing the pollen to the wind. The single carpel has two feathery styles for capturing airborne pollen. Upon fertilization, the single ovule develops into a caryopsis which, being light, is generally dispersed by the wind and, being buoyant too, sometimes by water. Whole plants are sometimes transported by the sea. The sugar cane on Cocos Keeling Island was derived from a clump from Java, 1,126km (700 miles) away, and bamboos have also been moved from place to place. In Europe, *Puccinellia maritima*, *Elymus arenarius* and *Ammophila arenaria* are dispersed by drifting rhizomes in rivers and the sea.

The vegetative shoot of the grass grows in a rather peculiar fashion. The very closely-noded stem constantly produces new leaves at its tip, thus producing a continuous sequence of growth, whilst remaining extremely short. New short-stemmed leafy shoots branch out from the axils of the leaves and these shoots in turn produce further shoots from their leafy axils, so that a large number of shoots can be produced without any marked lengthening of the stem. This process is called 'tillering' and is particularly noticeable in young cereal plants where it is a valuable asset in that less seed per hectare is needed to produce a good number of shoots.

Food-source for livestock

A typical grass turf, about 30cm (12in) high consists almost entirely of leaf, with the stems very short and hidden from view and lying, with the buds, within a centimetre of the ground. (Of course with exceptionally tall-growing grasses like the bamboos, the vegetative stems are much longer.) This characteristic is why grasses form such a valuable food for grazing animals, since in this vegetative state they can be repeatedly grazed without damaging the stems and buds, and a constant regrowth of leaves is ensured. Few other plants can survive such treatment. Of course, if grazing becomes over-intensive, and the grass is nibbled right down, the grass plants will become damaged. When flowering time arrives, the mature shoots elongate to bring the flowers up above the level of the leaves, tillering and leaf production ceases and, following the shedding of the fruit, the shoot dies. Shoots remaining in the vegetative state then further the growth of the plant.

Principal Cereals of the World

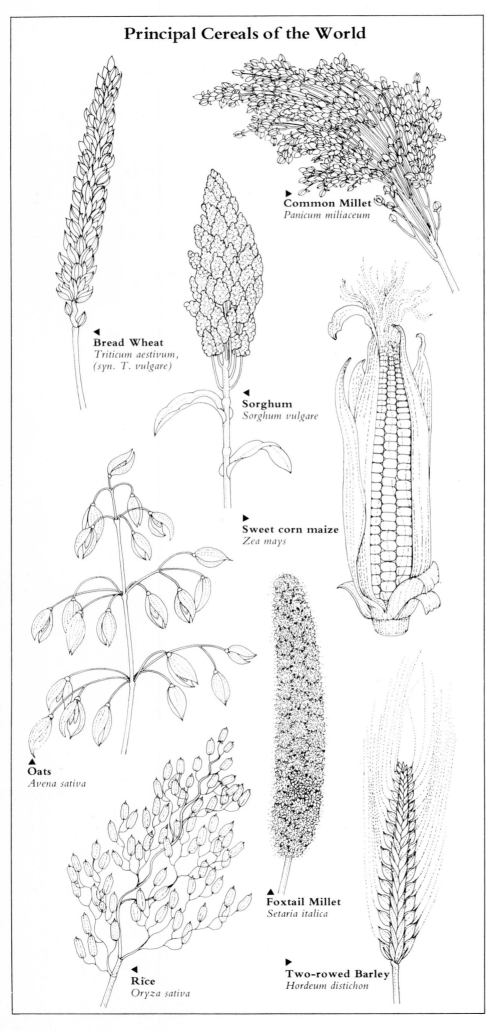

Common Millet
Panicum miliaceum

Bread Wheat
Triticum aestivum,
(syn. T. vulgare)

Sorghum
Sorghum vulgare

Sweet corn maize
Zea mays

Oats
Avena sativa

Foxtail Millet
Setaria italica

Rice
Oryza sativa

Two-rowed Barley
Hordeum distichon

This characteristic growth of grasses also permits mowing to provide winter fodder. Grass is at its peak in nutritional value when it is young and short, but mowing is usually left until the flowering heads appear, as prior to this it is difficult to handle the crop efficiently enough for natural drying (the cheapest method). At this stage too, the maximum weight of herbage per hectare is obtained. Any later and the plants become coarse and stemmy following the setting of the seed. It is important to have a leafy crop rather than a stemmy one because the leaves are more palatable and contain more protein than the stalks.

Species of grass used for feeding to livestock are grown for their vegetative parts; some other members of the grass family, the cereals, are cultivated for their grain.

Food-source for man

One or more of the cereals has formed the staple food for mankind in all the important civilizations of the world. They were originally derived from wild grasses and have arrived at their present highly cultivated state through ages of selective breeding. Some of these 'wild progenitors' still exist today, and so the hereditary pathways which our crops followed can still be traced.

The cereals are very adaptable plants, which is why they are cultivated all over the world. They tolerate a wide range of soils and climates and are straight forward to cultivate. Grain stores well in large quantities so that it can be carried over from one season to the next, and is easy to handle for transportation. Also cereals breed easily; most are self-pollinating, with the exception of rye. Most cereals are annuals, enabling harvesting in the same growing season as sowing. Some varieties are winter-hardy too.

If the three most important food plants in the world had to be picked, the lots would almost inevitably fall to three grasses: rice, wheat and corn (maize). Just over one-half of all the harvested land of the world is devoted to them and together they account for well over three-quarter billion tonnes of grain annually. The remainder of the world's crop of about one billion tonnes is made up of other grains: barley, oats, rye, millet and sorghum. Just over half the world's production of these grains comes from the USA, the USSR and western Europe.

Today's high-yield grains contain about 5–13 percent vegetable protein, which is rich compared to the only other staple crop that approaches them on a worldwide basis: potatoes. Wheat and rice together supply about 40 percent of man's food energy.

Rice

Half the world's population subsists wholly or partly on one grass—rice. This annual with its smooth narrow leaves and stems from 50–100cm (20–40in) long, has its origins as a swamp grass and may well have been the first crop to have been raised in the Far East. Today over 2,000 varieties of rice exist so that the crop may be cultivated in a wide range of soil and climate conditions. Rice grows best in tropical and subtropical flood plains and river valleys where the young shoots are transplanted from their seed beds into the 10–20cm (4–8in) water they need to attain maximum growth. When the grains in the upper spikelets are just ripe, the rice fields are drained and about two weeks later the crop is ready for harvesting. Following threshing, the grain, like barley and oats but unlike wheat, retains its husk. This is removed, giving brown rice; modern taste demands further milling (polishing) which strips the grain of most of its oil, protein, minerals and vitamins, leaving roughly 90–94 percent starch. Previously the disease beri-beri (weakness) was common amongst people relying heavily on rice for their food, but now the missing elements can be preserved in milling or added to the diet in other ways.

Wheat

Wheat (*Triticum*) is grown on nearly one-third of the land devoted to grain production. Unlike rice, it does not grow well in the tropics, where one of its major diseases, the wheat rust fungus, flourishes. Wheat is grown where summers are hot and dry, and winters cold and wet. It is one of the oldest of the cereals and many subspecies and varieties have been bred to suit different conditions.

Maize

Maize (*Zea mays*), or Indian corn, was first domesticated from the wild maize by the Indians of tropical America over 7,000 years ago. The cultivation of maize progressed and spread rapidly for the Indian was skilled in the art of plant breeding. Hybrid plants of more vigorous and prolific growth were produced by breeding maize with a wild grass relative, *Tripsacum dactyloides*, giving teosinte (*Euchlaena mexicana*), which was bred back with true maize, giving the hybrid plants. Today's maize cobs carry 500–1,000 kernels whereas the earliest known cultivated cobs bore only 48 kernels.

Cobs develop in the axils of the middle leaves of the main stem which is stout, solid and bears fleshy broad-bladed leaves. The female flowers, borne on separate inflorescences, possess long delicate pink or orchid-coloured styles called 'silks' which wither later as the ripening grains pass from their early soft 'milk' stage into the mature hardened state.

Maize is indeed a useful plant. Not only is the grain a popular vegetable, but it is ground into flour, used to make alcoholic drinks and sweet syrups, and, when dry, as poultry feed; leaves are used as fodder; the spathes surrounding the unopened spikes are used in paper-making; the dry cobs are used for firing. Even the stalks are used, as the fences, walls and roofs of whole villages in Guatemala can testify. In countries where amounts of sunshine are not sufficient to fully ripen the grain, maize provides green fodder or silage.

Other cereals

Other cereals important to man include: oats (*Avena sativa*), the ancestors of which originally occurred as weeds in other crops; rye (*Secale cereale*), which can survive soil acidity, poor hungry land and dry light soils; and barley (*Hordeum vulgare*), which can be grown as far north as the Arctic Circle and was formerly used for making bread and fermented drinks in the most ancient civilizations of the Old World until wheat replaced it.

Sugar cane

A grass prized for its sweet juice since ancient times is the sugar cane, *Saccharum officinarum*, which today accounts for two-thirds of the world's annual sugar production. Originating in India and New Guinea, it is a plant of tropical and subtropical countries, requiring a moist rich soil, preferably near the sea. It is in fact one of the few perennial grasses which has its vegetative parts used for human food, and unlike other grasses, it is selected for sterility. Many

canes never flower; those that do rarely produce fertile seeds, and the plant is normally propagated vegetatively. The plant grows to a height of 4–5m (13–16ft) in 8–15 months. At maturity, it has stout solid stems or canes, varying in diameter from 1–6cm (0.4–2.4in), from which the sugar is extracted by crushing. The residue, known as bagasse, is used in the manufacture of wallboards, plastics and paper, and in oil-refining. Molasses, another by-product, is used both as cattle feed and in chemical manufacture, including boot-blacking and rum.

Bamboo

The bamboos (Bambuseae) are the largest grasses. Their strong woody stems (culms) are hollow, jointed and glossy, and of variable colour. Strange legends surround many of the unusually-coloured specimens.

Bamboo shoots, with cells rich in protein, are amongst the fastest growing stems. At Kew Gardens, a culm of *Bambusa arundinacea* grew about 91cm (36in) in 24 hours! This is caused by a combination of active cell division and cell elongation in as much as one metre of the stem tip. Typically a height of 30m (98ft) can be attained within a matter of months, as compared to decades in the case of trees. In fact the individual bamboo cells do not grow any faster, there is just more of them doing the work.

The uses of bamboo for mankind are infinite: house, boat and bridge building, fishing and farming gear, musical instruments, paper, and food are just a few amongst them. They are generally natives of tropical and subtropical regions, where litter from the dense vegetation provides the rich deep loamy soil that bamboos prefer. However diversity amongst grass species is typical, and the bamboos are no exception; some species can survive at temperatures below freezing point and can be found on the snow-line of the Andes (4,500m—15,000ft), and at 3,000m (10,000ft) and more in the Himalayas.

Many species of bamboo are notably long-lived, and when they die it is in rather a strange fashion. In effect, they commit suicide every 33 to 66 years. At these times, the giant bamboos spontaneously burst into flower with great flowering spikes taking the place of the ordinary leafy shoots. In so doing, the plant uses up all its energy reserves without replenishing them in any way, and thus the plant dies. Strangely, all bamboos of the same species flower in the same year, wherever they may be growing. A Javanese bamboo transplanted to Jamaica will still flower at the same time as the rest of its Javanese relatives. Obviously, the year of the bamboo flower can be a calamitous one for the natives of tropical regions where a particular species is heavily relied upon to supply the bulk of local building materials.

Lawns

Many grasses make beautiful ornamental plants. Lawns are an outstanding example of the collective beauty of grasses. In establishing a lawn, however, only certain species can be used if the desired effect is to be obtained. For temperate climates, the genera *Agrostis* and *Festuca* provide the best lawn grasses, meeting the requirements of a dwarf habit with firm leaves and persistent growth to withstand regular hard cutting by a mowing machine. Of the two genera, *Agrostis* tends to be the more aggressive and enduring. Under normal conditions, on a well-drained soil, a mixture of 70 percent chewings fescue (*Festuca rubra* subsp. *commutata*) and 30 percent New Zealand browntop (*Agrostis tenuis*), is a popular lawn seed mixture.

*Below: Oats (**Avena**) are among the important temperate cereals, suitable especially for colder latitudes. Though very nutritious they are used primarily as animal fodder.*

AQUATICS
back to the water

The many thousands of plant species which today cover the land surface of the earth have all evolved from ancestral forms which were aquatic. Indeed, life itself had an aquatic origin. Land plants have thus adapted to life out of water and this is reflected in their structure and, especially, their water relations and their methods of reproduction. But if we look closely at the aquatic plants of today they show traces of features such as cuticles and stomata which are characteristic of land plants —why? It is now known that these are plants which are secondarily aquatic, that is to say they are descended from land plants and have subsequently invaded the water.

Aquatic plants are found in most of the major taxonomic groups, ranging from minute planktonic and surface-dwelling algae, through a relatively small number of species of mosses and ferns to the many species of flowering plants (angiosperms) which comprise most of the aquatic plants. There are no truly aquatic gymnosperms.

Apart from the algae, aquatics range in size from free-floating forms less than 1mm (0.04in) across to full trees, although most are herbaceous. The re-invasion of water has in many cases resulted in marked alteration and reduction in the morphology of the plants, so that it is sometimes hard to determine the taxonomic relationships of different species.

To sink or swim—types of water plants

If the main problems usually faced by land plants include those of support and lack of water, the environmental stresses on aquatic plants are nearly all due to 'too much' water. The depth of the water and the strength of the current will profoundly influence the type of plant that can grow at a given site. The physical density of plant tissue is quite close to that of water, so that plants that are fully immersed have few problems of support, unlike land plants living with their foliage in the air. The water plant must, however, be able to withstand the stresses imposed by the flow of water. In most cases this is achieved by having highly flexible stems and leaves which will readily bend into a shape that offers least resistance to the water flowing by, but which are strong enough to resist the pull of the water.

The degree to which the plant is immersed in the water will obviously influence the effects of water flow, and water plants can be divided into three groups: submerged, floating and emergent, depending on the position of their leaves relative to the water surface.

Submerged water plants
Submerged water plants are those whose vegetative parts are normally beneath the water at all times (although most still produce aerial flowers). Virtually all of these plants are rooted in the substrate, although a few, such as the horn-wort, *Ceratophyllum,* have no roots and float freely below the surface. Submerged plants usually have one of two growth forms. Firstly, as a rosette with the leaves all arising from a short rhizome or root stock, as in the shore-weed, *Littorella,* the cape pondweed, *Aponogeton,* or in the quill-wort, *Isoetes,* a distant relative of the ferns. In the second type the leaves are borne along long flexible stems. Some of these may root in several places. Typical of this group are the pondweed, *Potamogeton,* and the Canadian pondweed, *Elodea canadensis.* A similar growth form is found in submerged aquatic mosses such as *Fontinalis.*

The leaves of submerged plants are frequently elongated or linear in shape. *Potamogeton compressus* has leaves 10–20cm (4–8in) long by only 2–4mm (0.08–0.16in) wide, whilst in *Elodea* the leaves may measure 10mm by 2–3mm (0.4 by 0.08–0.12in). Less elongated submerged leaves are often oval or lance shaped, as in *P. lucens.* Dissection of submerged leaves occurs in many species, such as the water milfoil, *Myriophyllum verticillatum,* and the water crow-foots, species of *Ranunculus.* Here, the leaves, which may be opposite or whorled in their arrangement on the stem, are divided into numerous thin strands, sometimes as many as 40 or 50. The advantage to the plant of linear or dissected leaves probably lies in the fact that

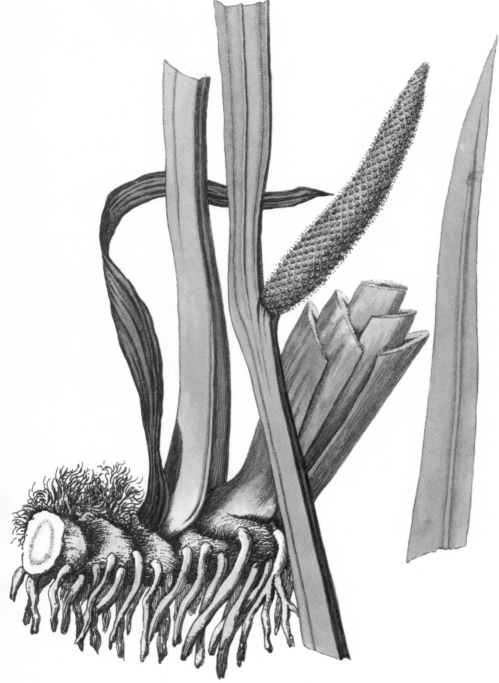

Acorus calamus
Sweet Flag

Above: An amazing assemblage of floating plants on an Indian lake. Among the large rounded water lily leaves are the smaller ones of fringed water lily, the rosettes of water chestnut and the paired leaves of the aquatic fern Salvinia.

they will offer relatively little resistance to flowing water and are less likely to tear, compared with broad leaves.

The internal structure of submerged leaves and stems reflects the nature of the aquatic environment. With little need for support there are few fibres or strengthening tissues within the stems and few woody, or lignified, vessels in the xylem. In many submerged plants the whole vascular system, especially the xylem, is markedly reduced compared with that in land plants, sometimes with the complete loss of xylem vessels and their replacement, as in *Potamogeton*, by a cavity.

Another striking feature is the presence of many cavities, or lacunae, within the tissues. These elongated cavities are found in roots, stems and leaves and in some species may occupy the bulk of the tissue volume. Their number and arrangement are often characteristic of a particular genus or species. For example, *Isoetes* has just four large lacunae running along each leaf, whereas the stem of *Myriophyllum* has a ring of lacunae around the central vascular strand and the leaves of *Potamogeton* appear to consist of a network of lacunae when viewed in cross-section.

Perhaps the most extraordinary group of submerged plants are those which belong to the family Podostemaceae. They are all tropical or subtropical and their structure is so reduced as to form a thallus with no true stem or roots. The thallus, which is sometimes branched, may be floating or attached to rock surfaces and in some species bears a variety of lobed or dissected leaf-like structures. Although clearly flowering plants the evolutionary relationships of this family are most obscure.

Floating water plants

Whilst most submerged plants may remain more or less hidden beneath the water those water plants whose leaves float on the surface are more often noticeable due either to the beauty of their flowers, as with the water lilies, or to their presence as disruptive weeds blocking waterways and lakes, as with *Salvinia* and *Pistia*.

The environment of a floating-leaved plant is in some ways even more specialized than that of the submerged plant. While the lower surface of the leaf is immersed in water, the upper surface is exposed to the rigours of the aerial environment. The effects of wind and waves will be particularly pronounced, wind tending to lift and tear the leaves while waves may tear or swamp them. It is thus not surprising that floating vegetation tends to be found on relatively sheltered waters. Floating leaves are often oval, as in *Potamogeton natans*, or near-circular in outline, as in the water lilies, *Nuphar* and *Nymphaea*. A circular leaf form may be less likely to be swamped by waves and less likely to tear. The extreme example of this form is shown by the giant Amazonian water lily, *Victoria amazonica,* whose leaves may reach 1.5m (5ft) in diameter and possess an upturned rim more than 15cm (6in) high. The huge leaves are reinforced underneath with a system of stiff ridges and are so buoyant that in the mid-nineteenth century Joseph Paxton was able to float his young daughter on such a leaf!

When water does fall on the leaf surface, from waves or rain, its displacement is often aided by the presence of a waxy water-repellent cuticle on the upper leaf surface. This cuticle will also be important in reducing the evaporation of water from the leaf tissues, and in this respect the upper surface of many floating leaves is similar to that of land plants, with functioning

Far right: Most floating leaves like those of water lilies (Nymphaea) are round to oval. It is thought that this form is less likely to be swamped or to tear in rough water.

Right: Mangroves belong to three different families which have become adapted in similar ways to living in tidal mud flats. This is Avicennia resinifera in New Zealand.

Below: At low water the pneumatophores or breathing roots of the Avicennia are seen entirely covering the exposed mud. They contain apertures so that air can reach the roots below.

stomata. Unlike land plants, however, chloroplasts are often found in the epidermal cells of the upper surface. The lower leaf surface is usually without stomata, although relict non-functional stomata may be present in some plants, such as *Potamogeton*. The cuticle on the lower surface is often absent or relatively thin. Lacunae, like those of submerged plants, are found in many floating leaves, and being gas-filled are important in maintaining buoyancy.

Where floating-leaved plants are rooted on the bottom, the leaf stalks (petioles) usually have a structure similar to that of submerged plants. Rooted plants such as these are rarely found in deep water, although the petioles may be long and trailing. This is especially important if the plant is growing where there may be changes in the depth of water. If the petioles were too short and the water level rose the plant could 'drown'.

Plants which are free-floating are again usually found on relatively slow-moving and sheltered waters. In such situations they may become extremely abundant, spread rapidly and become weeds of international importance.

Unlike the rooted floating-leaved plants those which are free-floating show a great variety of structure. Some of the simplest are the duck-weeds, *Lemna*. This genus of worldwide distribution has no separation into stem and leaves, but merely consists of one or more plate-like thalli usually less than 1cm (0.4in) across. The thallus, which bears a single root, is kept buoyant by numerous lacunae. Even smaller is another member of the same family, *Wolffia*, the smallest of all flowering plants, a tiny globular thallus with no leaves, shoots or roots, often less than 0.5mm (0.02in) across. Even so, *Wolffia*, like *Lemna*, may at times entirely cover the surface of sheltered water bodies.

Many free-floating plants have a rosette type of growth form; the leaves are borne in whorls or spirals on a form of shortened central stem. This is well shown in *Pistia stratiotes*, an important weed of tropical waterways. The rounded leaves, 1–2cm (0.4–0.8in) across, have a deep central 'keel' composed of spongy tissue well supplied with lacunae which acts as a float. The upper surfaces of the leaves are covered with dense hairs which are water repellent and prevent the leaf from becoming swamped.

Water-repellent hairs are also found on the upper leaf surfaces of the infamous water-fern *Salvinia*. This plant bears its leaves, 2–5cm (0.8–2in) long, in pairs along a short stem. Buoyancy is maintained by air trapped between the hairs on the leaf surface. There are no roots, but much-branched feathery modified leaves hang down in the water below the floating leaves and carry out the same functions as roots. The effectiveness of the flotation mechanism in *Salvinia* was demonstrated in Central Africa when Lake Kariba was formed soon after 1960 by the damming of the River Zambezi. As soon as the lake began to fill extensive colonies of *Salvinia* appeared on its surface. Growth was extremely rapid, resulting in hundreds of square kilometres of the lake surface being covered by dense mats. The only way *Salvinia* could have reached Lake Kariba was from swamps upstream on the Zambezi. The Victoria Falls are *en route*, and *Salvinia* reaching

Left: The water hyacinth, Eichhornia crassipes, floats by means of air-filled leaf-bases. Originating in tropical America, it has become a major pest of tropical waterways.

Kariba must have survived a drop of over 100m (328ft)!

Originating in Central and South America, *Salvinia* has since the 1940s become a serious weed, especially in Central Africa and Ceylon. The dense rapidly growing mats of vegetation choke the waterways, making fishing and the passage of small boats virtually impossible. As the old dead leaves decay the oxygen in the water beneath may become depleted, to the detriment of any animal life such as fish originally present.

Emergent vegetation

Emergent vegetation, the third type of aquatic plant, may often form extensive stands in relatively shallow water. These are plants which are rooted in the substrate below the water and whose foliage is borne wholly or in part in the air above the water. Structurally this emergent foliage is often little different from that of land plants, with well-developed vascular tissues and with cuticle and stomata on both sides of the leaves. Lacunae may still be present in some species, however, and it should be remembered that the roots, stem bases and young shoots will still have to exist in an aquatic environment.

Emergent plants may be dicotyledons, such as the bog bean, *Menyanthes trifoliata*, or monocotyledons, such as the common reed, *Phragmites* or the reedmace, *Typha*. Monocotyledons often form conspicuous dense stands of vegetation at the edges of rivers and lakes, forming a community known as reedswamp. The combined effects of the plant shoots slowing the water currents, causing the deposition of silt, and the accumulation of dead plant remains may lead to the gradual filling of the water body, leading to the formation of drier land bearing different plant communities—a type of succession known as a hydrosere.

The productivity of many reedswamp communities is high, the stands often having a large total leaf area but allowing an efficient distribution of light down to the lower leaves. In some parts of the world reedswamp plants are of economic importance, as in Roumania, where the vast areas of *Phragmites* in the Danube delta are the basis of a major paper and chemicals industry.

One of the first water plants to be important economically was papyrus, *Cyperus papyrus*, which was used in Ancient Egypt for building and paper-making, and which is still important locally. Although normally a rooted emergent, papyrus may at times form floating rafts upon which other plants may become established. In Central Africa these rafts are known as Sudd and may form vast areas of virtually impenetrable marshy thickets.

Too much and too little—the aquatic environment

The degree of light penetration down into the water will obviously influence the depth to which submerged plants can grow, the limit usually being at intensities of between 1 and 3 percent of full sunlight. The depth at which this occurs will depend on the amount of silt and coloured material in the water and on the degree of shading due to aquatic plants and plankton. In some rich or silty lakes the limit to growth may be at a depth of 1m (3.3ft) or less, whereas in some clear, unproductive lakes submerged plants may be found down to depths of as much as 10m (33ft).

Oxygen diffuses more than a thousand times more slowly in water than in air, so the supply of oxygen to the submerged parts of water plants may be limited. Indeed, the roots may often be in a virtually oxygenless environment. The presence of lacunae and air spaces in floating and emergent plants will provide a pathway allowing oxygen to diffuse to the submerged parts. It has been shown, for example, that oxygen may diffuse out from the submerged root tips of *Menyanthes*. Submerged plants have no access to the air, but oxygen deficiencies may be partially relieved by the retention in the lacunae of some of the oxygen released in photosynthesis. This may, in fact, be one of the ways in which lacunae are formed. Nevertheless, many water plants undoubtedly undergo periods when their growth is limited by lack of oxygen.

Mineral ions are taken up by water plants by both the roots and the submerged foliage, and growth may be critically affected by the concentrations of the minerals in the water. In particular most water plants are intolerant of the saline conditions found in salt lakes and estuaries. In such conditions high salt concentrations may disrupt the water balance of the plant or lead to the accumulation within the tissues of harmful amounts of salts, as well as affecting the uptake of necessary minerals. Some plants, such as *Phragmites communis*, can tolerate a wide range of salinities, being found in both freshwater and in estuaries. Others, such as *Scirpus maritimus*, are nearly always found in brackish or estuarine conditions. Partial immersion in full sea water is tolerated by the trees of the tropical mangrove swamps. The mangrove roots are often in highly oxygen-deficient muds, and have specialized roots, pneumatophores, which project above the mud and water surface enabling air to pass into the root system.

Only a few flowering plants have become adapted to life totally immersed in the sea. These plants, such as *Posidonia* and *Zostera*, the eelgrass, flower and spread by seed in the sea—true marine flowers.

Reproduction in the water

In many water plants reproduction is usually vegetative. This is especially true in free-floating plants where their rapid spread could not occur fast enough by means of purely sexual reproduction. Some plants such as *Lemna* reproduce simply by budding of the thallus, whilst others, such as *Pistia*, form stolons, spreading horizontal stems on which new plants are formed. Propagation by the simple break-up of the old stems may also occur, as in *Salvinia*. Vegetative reproduction in rooted plants is often by means of tubers or by underground stems, rhizomes, as in *Phragmites* and papyrus.

Sexual reproduction in aquatic plants is in most cases similar to that in land plants, especially in floating and emergent plants. Many submerged plants, too, have not entirely lost their links with the air, putting up flower bearing shoots above the water, as in the bladderwort, *Utricularia*. These aerial flowers are usually either wind- or insect-pollinated. Some water plants may reduce the complexity of the flowers with, for example loss of sepals and petals. *Wolffia* shows an extreme example of this, the male flower consisting of a single stamen only.

Only a few genera, such as *Ceratophyllum* and the marine *Zostera*, have become true water plants in that they flower below the surface. In *Ceratophyllum* the anthers from the male flower have a small float. When mature they detach and float to the surface where they dehisce, releasing the pollen. The pollen grains sink through the water, and in doing so some of them may come into contact with the elongated stigmas of the female flowers. Thus with water pollination true independence of the air is achieved and the re-invasion of the water is complete.

*Right: Water buttercups like **Ranunculus peltatus** have both floating and submerged leaves when they grow in still water. The long underwater stems produce masses of flowers.*

*Below: Species of **Cabomba**, related to the water lily, are often grown to oxygenate aquaria. The finely cut submerged leaves are replaced by broad floating leaves on the surface.*

CLIMBERS
plants that cheat

Plants grow towards light. This is because they need the energy from light to make their food. If one plant grows faster and taller than another, its leaves may overshadow those of the smaller plant. This will stop the smaller plant getting vital light. It will grow more slowly, become more shaded and possibly die from light starvation.

To grow upwards, become tall and bear many leaves, a plant must have a strong stem. It must be structurally sound and so a plant grows wider at the base as it becomes taller. It produces cells which elongate and become cylinders of wood. Cylinders are mechanically strong. They are produced in a ring around the inside of the stem giving a cylinder of wooden cylinders within the stem.

Climbing plants cheat. They avoid using their food and energy to build supportive wood. Instead, they use some of their total energy to produce structures or organs to attach themselves to another plant or surface which leads upwards. The food and energy they have saved can be used to grow faster, further and higher than neighbouring plants. In this way they stay in the light and gain even more energy.

The reduction in supportive tissue, the rapid growth rate and production of large amounts of food materials is reflected in the internal construction, or anatomy, of the stems of climbing plants. There are relatively little amounts of the woody tissues called xylem and sclerenchyma but there are large amounts of the food transport tissue, phloem. In most plants a collection of water-conducting pipes and a collection of food-conducting pipes will be grouped together as a bundle, called a vascular bundle. In many climbing plants, for example marrows and cucumbers, there is an extra collection of food pipes so that the xylem is sandwiched between two layers of phloem.

Those climbers which become woody do not have wood as a solid cylinder pierced by small rays of conducting tissue. The wood runs the length of the stem in wedge shaped layers separated from each other by equally large wedges of food-conducting tissue. This may be seen in the stems of the vine, *Vitis,* and the Dutchman's pipe, *Aristolochia.*

The redirection of energy into the production of climbing aids has given a variety of specialized structures, organs and ways of using them. It is possible to artificially arrange climbing plants into four groups on their modes of climbing.

Penetrating climbers

Penetrating climbers are mainly found within the genus *Hedera,* ivy. Special roots grow out from the cable-like stem. These roots can grow into the surface of a wall or the bark of a tree, anchoring the plant to it and enabling it to grow up the surface. The connection to the original root may be severed, but the plant can survive by absorbing water and nutrients from the surface covered, though they are not in any sense parasites.

The genus *Hedera,* is found throughout Europe and Asia, north of the tropics and sub-tropics, and even in the Canary Islands as *Hedera canariensis.* This is an endemic species much favoured by gardeners for its ornamental variegated foliage.

Falling climbers

Falling climbers make no positive attempt to climb nor do they attach themselves to any other plant or structure. They grow upwards until they fall over onto vegetation which they then grow through and over. Such plants are also called scramblers, stragglers or trailing plants. Their main specialization is in the form of grappling hooks or barbs growing from their outer layer or epidermis. Examples are roses and brambles of the family Rosaceae. The 'prickles' on their leaves and stems are outgrowths from cells of the epidermis. When they are hook-shaped the barb points back in the direction from which the shoot has come. If the stem should slip backwards the barbs may catch in the vegetation and check the slip. *Galium,* or bedstraw, has a similar mechanism. Numerous small prickles cover the stem and leaves of the plant, making it a very efficient straggler.

The genus *Drusa* is a rather peculiar member of the family Umbelliferae. Its flowers are not arranged in an 'umbel' but are solitary, and it has a very odd distribution, being found in the Canary Islands and South America. It is a straggling plant with numerous fragile epidermal 'prickles' on its stems, leaves and fruits. These epidermal structures also secrete sticky substances which help the plant maintain contact with its support.

Bryonia dioica
White Bryony

Many climbing plants produce spirally coiling tendrils with which they attach themselves to other growth on their way upwards, as in this tropical cucumber relation **Momordica.**

Twining climbers

Twining climbers curl around a support and grow up it. The mechanical strength of their spiral, and the friction of their surface contact with the support, stop them from sliding down. The mechanism by which this occurs consists of two stages.

The first stage is that before contact with a support is made. The shoot of the plant grows upwards or sideways in a helical fashion. It does not grow in a straight line, but moves round at varying angles as it grows. This may be related to the way the cells of the stem are produced spirally at its apex.

Cells are not produced all together as a single mass. Cells are formed in a spiral around the central dome of the apex. They will therefore be at different stages in their development at any one time. Once produced, their development, differentiation and growth will be under the delicate and complex control of many biochemical compounds, including plant hormones. Some of these enable the cellulose walls of young plant cells to become more 'plastic'. Increased water pressure inside the cells causes them to elongate. This can give rapid elongation of the stem (or root). It is possible that the pattern of cell production and their subsequent differences in development at the time of elongation causes uneven elongation to occur, giving the circling motion known as circumnutation. All plants show circum-nutation to some extent, though in many twining plants it is very marked. *Humulus,* the plant which gives hops used in brewing, circum-nutates in a clockwise direction, while convolvulus, the bindweed, circum-nutates in an anti-clockwise direction. Once a support has been touched by the circum-nutating shoot, the second stage of the twining process begins.

This second stage is dependent on the stimulus of touch on the surface of the shoot. It would seem that hormonal control of growth is again involved, but in this case as a direct response to an external stimulus. The surface of the shoot furthest *away* from the touch stimulus grows longer more rapidly than the contact surface. This suggests that the touch stimulus has slowed down, or inhibited, hormone production at the contact surface. It causes the shoot to bend inwards towards the contact and grow around the support. Once begun this twining, or coiling, pattern of growth is self-perpetuating. Growth in response to touch is called thigmotropism.

Plants maintain the same direction during twining as they show in circum-nutation, hop and honeysuckle twining clockwise and bindweed anti-clockwise. Other twining plants are *Clematis, Bryonia* (bryony), *Aristolochia* (Dutchman's pipe) and the popular ornamental *Stephanotis floribunda.*

Grasping climbers

The pattern of growth in grasping climbers is similar to that of twining climbers. There is an initial growth phase during which circumnutation may occur, then after contact has been made thigmotropic coiling around the support occurs. The essential difference between this group and the twiners is that the grasping plants have separate organs specialized for the grasping and coiling. These organs are known as tendrils. They are very common in the legume family Leguminosae, and the gourd family Cucurbitaceae. The pea, *Pisum sativum,*

Ivy climbs up trees to obtain maximum light for flowering and fruiting; it clings by aerial roots. It is not a parasite but a strong ivy plant can eventually choke elderly trees.

and the sweet pea, *Lathyrus odoratus,* are legumes, while bryony, *Bryonia cretica,* and cucumber, *Cucurbita,* belong to the gourd family.

Tendrils often develop instead of the end leaflet of a pinnate leaf, or in place of any or all of the leaflets. *Lathyrus aphaca,* the yellow vetchling, has tendrils instead of leaves. Photosynthesis is carried out by the greatly enlarged stipules.

During the circum-nutatory or 'searching' phase the tendrils remain mechanically weak. This keeps them flexible though fragile. Support is mainly due to water pressure inside the cells. However, once contact has been made and the tendril has coiled around the support, woody strengthening tissue may be laid down. The tendril continues to coil around the support and along its own length; so that the plant and support are pulled closer together. This coiling gives the tendril extra strength while keeping it flexible enabling it to remain intact on windy days when violent movements may occur. Not only leaflets can develop as tendrils; leaf-stalks, leaflet-stalks and elongated leaf mid-ribs may act in a tendril-like fashion. The Chinese *Parthenocissus tricuspidata (Ampelopsis veitchii)* even has tendrils which develop sticky pads at their ends. The plant does not grasp; it sticks instead.

The climbing and straggling habit is widespread throughout most of the climatic areas of the world. It is generally a weedy feature used for the exploitation of other plants and is frequent among colonizers of disturbed environments. As many of the climbers are good biological weeds, they grow rapidly and flower profusely. This has made them popular ornamental plants. Unchecked they can rapidly become unpopular horticultural weeds.

ALPINES
life at the top

Alpine plants are so-called because they live in the alpine zone of mountains. In this area which lies between the timber-line and the zone of permanent snow fields where plant life is absent, dwarf shrubs and 'cushion plants' predominate in alpine meadows, on scree and in crevices of rocks. These plants live in the face of a climate so extreme that evolution has moulded their forms in a highly specialized and characteristic fashion to survive in conditions as harsh as may be found anywhere on the globe. Floods of rain, drought, baking heat, extreme cold, winds of intense ferocity, radiation and unstable soil conditions all have to be withstood if a plant is to survive at such levels of exposure. Any slight rise in the terrain which may afford shelter from the biting wind or temporary protective blanketing of snow in a hollow are used to maximum advantage in this inhospitable environment.

The alpine zone is typically found above around 2,400m (7,900ft) in central and southern European mountain systems, higher than this in mountains of warmer climes. In the East African mountain ranges, for instance, the alpine zone occurs between 4,000m and 5,000m (13,000ft and 16,500ft), while a typical mountain of tropical Indonesia or Malaysia has alpine shrubs above 3,800m (12,500ft) and alpine herbs above 4,500m (14,800ft). Here, at these upper limits of plant growth, the land is rock-littered and rolling, broken by peaks and cliffs, with stretches of bare wind-blown rock faces interspersed with snow pockets, with patches of brown vegetation, either dead or dormant. Wind whistles across the scene perpetually, sometimes at devastating speeds of 160kmph (100mph) or more. Despite such difficult conditions for growth and reproduction, some of the most perfect and delicate blooms of all the flowering plants are found in this zone. The tiny exquisite snow buttercup, *Ranunculus nivalis*, has been found budding under a 3m (10ft) snow bank, whilst the little bell-bloomed stalks of the alpine soldanella, *Soldanella alpina*, which melt their way up through their snow covering by radiating stored energy as heat, contrast sharply with the severity of their surroundings. Spring and summer, though of short duration, can transform the scene into a blaze of flowering colour.

Growth and shape

Survival at these heights depends largely on the adoption of one particular habit—low growth. The high winds of the alpine zone are often charged with rock particles and ice crystals; exposure to such winds invites damage to the delicate tissues of plants. Standing on an exposed mountainside, a man can be blown over by the force of the wind. Should he now sit down, he finds the effects of the wind's buffeting are considerably reduced. If he lies flat on the ground, he would find that the wind has little or no effect upon him now. Alpines are often called 'belly plants' by scientists because they grow so low, out of the wind's way, that an observer must lie on his stomach to investigate them. This characteristic habit is often reflected in their names—*acaulis* (without stem), *prostrata* (prostrate), *procumbens* (leaning forward), *humilis* (humble). The drag effect on the wind, which operates close to the ground, is increased by the presence of lumps and bumps in the surface which break up the flow of air, providing pockets of shelter from the wind. Even the tiniest irregularity can provide shelter for a plant which is small enough to grow within its lee.

As well as low growth, plants need to be supple so that they can bend before the blast. Rigid woody structures would soon be snapped off, and the only woody plants in alpine zones have pliable branches which can adopt all manner of shapes. The Rocky Mountain snow willow is such a plant. Though it hugs the

Ramonda nathaliae

*Right: The brilliant blue of alpine gentians such as **Gentiana clusii** attracts pollinating bees; inside the flower tube whitish markings encourage the insects to penetrate.*

ground, rising to only a few centimetres in height, it still manages to produce a profusion of catkins.

Cushion-shaped plants, inconspicuous ground-hugging domes, are another common feature of the alpines. Here closely-matted stalks clump together, providing an ideal form for cutting wind resistance, and also acting as a heat-trap. Thus despite great diurnal fluctuation of temperature externally, within the cushion a buffering effect is experienced as heat from the sun during the day-time is absorbed and retained. Temperature measurements inside a cushion of *Silene acaulis,* the cushion pink or moss campion (a numerous and successful alpine of the subarctic, found in North America, Europe and Scandinavia), have shown that its interior may be as much as 10°C (50°F) warmer than the surrounding air.

Many mountain plants occupy rock crevices. These chasmophytes, as they are called, escape competition from other plants, and the reach of grazing animals and man; the chasmophytic habitat can, in fact, provide a last refuge for plants under pressure from adversely changing.

environmental conditions and the activities of man. It is remarkable how their roots can obtain enough moisture and nourishment in the tiniest cracks in a rock or cliff.

Alpines grow extremely slowly. The addition of less than a centimetre of stem during a single season is quite normal in mosses and higher plants, lichens grow even more slowly. A typical cushion-plant seedling puts out as few as two minute leaves a year; a mat of mountain avens, *Dryas octopetala,* a metre across, was probably a seedling a century ago.

Reproduction

The little growth that these plants are capable of is concentrated in their short growing season into producing the flowers they need to reproduce and into the long roots which they need to collect moisture in the thin dry soils of these upper altitudes. Timing is all-important for alpine plants. During cold weather they lie dormant, conserving their energy, waiting for the summer season and the chance to reproduce. The arrival of the warmer weather heralds a

burst of activity, as the plants race against time to produce their flowers and seeds—in Norway, a snow buttercup, *Ranunculus nivalis* was found to flower five days after the snow had melted off the plant and only seventeen days later, it bore ripe seeds. But with summers so short, a seed is unable to mature and produce seeds of its own all in a single season, so most mountain plants are perennials, gradually building up their store of energy until it is sufficient for flower production. Of more than 300 species of flowering plants to be found above the timber-line in the Rockies, for example, only two are annuals, and one of these produces flowers of only pinhead size. Many mountain plants must wait for three or four years before they can risk blooming. The glacier lily, which waits seven years before flowering, can die in the struggle to replace its leaves should an animal eat them early in the season, and any careless picking by climbers almost always spells death. A mat of *Silene acaulis* may take as long as ten years to establish itself sufficiently to produce its first flower, and twenty years before full bloom is achieved.

How alpine cushions form

The cushion is a characteristic growth form of alpine plants: the tightly matted stalks close to the ground provide virtually no obstacle to wind, and also form a heat-trap inside which the sun's heat is retained at night. Cushions arise from a central root from which branches radiate, each bearing small leafy shoots packed together in the smallest possible space.

Many alpine flowers are brilliantly coloured, such as the rich wine-colouring of the purple saxifrage *Saxifraga oppositifolia*, the deep golden yellow of the snow buttercup *Ranunculus nivalis*, and the vivid dark blue of the gentian, *Gentiana nivalis*. These intense colours absorb light and heat, and are possibly particularly attractive to insects, of which there are a surprising number in view of the rigours of the environment and tendency to get blown away by the wind (60 percent of insects found above the timber line are actually wingless, and many of those which do have wings seldom use them). There are many pale-coloured and white flowers also; possibly these stand out particularly well against the dark green heat-absorbent leaves.

The warmth-retaining interiors of the cushions like *Silene acaulis* are a haven to cold-gripped insects which, whilst crawling around within, effect pollination. Many mountain plants are adapted for wind-pollination, an

*Left: The name saxifrage means 'rock-breaker' and many species, like **Saxifraga crustata**, grow in crevices. Rosette formation ensures that each leaf receives as much light as possible without spreading upwards.*

effective method in this zone of low plant growth and high wind velocities. Others are self-fertilizing, for example the grass *Festuca ovina*, while some species such as the mountain buttercup, *Ranunculus montana*, can set seeds without being fertilized at all. The mountain grass *Poa alpina* has seeds which germinate while still attached to the parent plant, while some other grasses produce new shoots instead of seeds, as with *Festuca vivipara*. The drooping saxifrage *Saxifraga cernua* forms bulbils which drop off and winter beneath the snow, producing new plants the following spring.

Structural and physiological adaptations

Much of an alpine plant's energy is directed towards its roots. Thin dry soils mean a wide root spread to obtain sufficient water, and blasting winds together with an unstable soil demand good root anchorage. Megarrhiza (large bulb or root) is a name which can be found associated with a number of alpine plants; a moss campion only a few centimetres high can have roots of a metre or more in depth. Long tap roots can often be found associated with plants growing in moving scree, where they

act as anchors. Such plants, slithering steadily downhill with the superficial surface, can often be found with the point of origin of the root actually above the plant!

Leaves show adaptations to mountain conditions, many alpines having adopted the 'evergreen habit' with one set of hard-earned leaves used for more than one season. Leaves are often thick and waxy, or furry as in the Swiss edelweiss, *Leontopodium alpinum*, to cut down on surface evaporation. Hairs also serve to retain heat; the snow willow *Salix repens* has fuzz-covered buds, the hairs being white and attached to a black core, so that light and heat penetrating the translucent hairs are absorbed by the black interior and retained there as escape is prevented by reflection. The willow bud acts, in effect, like a miniature greenhouse.

A number of alpine plants are protected from frost damage by having a rich cell fluid which acts like anti-freeze, for example the mountain crowfoot *Ranunculus glacialis*. Some plants are so well adapted to cold and a minimum of sunlight that they are able to start growing while still lying deep beneath a snow covering as with *Ranunculus nivalis* and *Soldanella alpina*. Snow is, in fact, a great protector and insulator; extreme temperature variations are not felt within

it and, provided it melts sufficiently at one point to allow sunshine to penetrate and the plant to photosynthesize sufficient starch to carry it through the winter, it can successfully nurture plant life.

Zonation of vegetation

At any latitude, as one goes up a mountain, provided that the vertical rise is sufficient, there is a zonation of climates which reflects the major climatic zonation throughout the world from the equator to the poles. Correspondingly, there is a zonation of vegetation belts as one rises up the mountain-side, so that a mountain compresses into a short space conditions and resulting vegetation types which on the flat would be spaced out over thousands of miles. Obviously there are local variations depending on the situation of the mountain. In certain tropical mountains, rain forest gives way to various levels of montane forest vegetation with progressively lower canopy levels, above which come heaths of tussocky grasses and sedges with gigantic woody tree-like forms of genera which in temperate climes are small herbaceous plants. African examples of this gigantism include the tree *Senecio*, a gigantic member of the groundsel family which grows to a rosette of cabbage-like leaves and a flowering spike, and giant lobelias which can be as much as 6m (20ft) tall. Gnarled heather trees occur too, draped in lichens and drooping over beds of mosses and liverworts. This phenomenon of gigantism occurs both in equatorial African and South

American mountain systems, though the plants involved are of different families and genera. Above this zone comes the alpine zone of low-growing flowering plants and lichens, and finally the zone of permanent snow and ice is reached.

Despite minor local variation, the harsh environmental factors which a plant must face at high altitudes are the same the world over, and as a consequence widely-separated plant species, which are quite un-related to one another and whose lowland-dwelling relatives do not look at all alike, have evolved into similar forms as adaptations to similar environmental pressures. This is true of many of the un-related alpine plants which have adopted the widespread cushion form. Such 'convergent evolution' is responsible for the strong parallels which can be drawn between the high altitude flora of two very separate mountain systems, for example the Andes and the East African ranges. Convergent evolution from such dissimilar origins is not hard to understand when one remembers that evolution is accelerated by spontaneous mutations which are more likely to occur under the following conditions: extremes of heat or cold, radiation, oxygen deficiency. All these factors are present at high altitudes and thus the higher one goes, the greater the potential for genetic change will be.

Alpine plants also show another noteworthy feature: many of them also occur in arctic regions north of the tree boundary. Plants such as the snow gentian, *Gentiana nivalis,* the mountain avens, *Dryas octopetala,* the rockfoil

*Above: Scabweed is the apt name given to the very dense, almost flat mats of **Raoulia australis,** a New Zealand composite with minute flowers. It grows in dry, rocky places.*

*Above right: Many primrose relations flourish in the Alps, like **Primula hirsuta** which grows up to 3,000m (11,000ft) altitude and has a special liking for bare rock, rooting into tiny crevices.*

*Right: Although many alpines are specially adapted for difficult conditions, some are normal herbaceous plants like **Aquilegia fragrans,** here seen beside a lake in Kashmir at nearly 4,000m (13,000ft).*

Saxifraga stellaris and the moss campion, *Silene acaulis,* belong in this arctic–alpine category. The Rockies share, in fact, as many as 65 species of flowering plants with the North American arctic. It is thought that this disjunct distribution arose when advancing ice forced the arctic flora southwards, and upon its retreat certain of the returning arctic species encountering mountain ranges became established at high altitudes, finding conditions there as suitable for their particular adaptations as farther north on the flat. Both environments are comparable in their severity; in both the survival of living things is poised on a knife-edge, a delicate balance between death and survival.

CARNIVOROUS, PARASITIC AND SYMBIOTIC PLANTS

killers and fellow travellers

Animals are in general regarded as active because of their need to obtain other living organisms for food. There are, however, certain plants which have abandoned the independence afforded by photosynthesis and the virtually total dependence upon a root system by which to obtain water and mineral nutrients, and obtain at least some of their needs from other living organisms but without abandoning the sedentary way of life so characteristic of plants.

Epiphytes

There are certain habitats which are, by their very nature, nutrient poor. Epiphytic plants, those which perch high on the branches of

*Drosera anglica D. rotundifolia
D. intermedia (left to right)*
Sundews

forest trees, certainly obtain more light for photosynthesis than their forest-floor cousins but they suffer the grave disadvantage of having little or no soil from which to get nutrient minerals and water. Some of the epiphytic orchids of tropical rain forests have overcome this problem by the use of aerial roots which absorb moisture directly from the humid atmosphere around them. Plants such as *Nepenthes*, the pitcher plant, have adopted a different approach to the problem.

Species of *Nepenthes*, whose range extends from Madagascar to North Australia, are to be found in wet jungles, growing either rooted in swampy soil or as epiphytes. The former species may send climbing shoots from their rhizomes in the soil to the tops of trees 15m (50ft) or more high. The basal part of each leaf of a pitcher plant extends into a tendril and ends in a vertically held pitcher. The pitchers range in size, depending upon the species, from structures that can hold 10 millilitres (0.02 pint) of liquid if filled to capacity to ones which can hold well over 1 litre (1.75 pint). The pitchers do contain liquid but only to about one-third of their total capacity and it is this pitcher liquid which holds the key to the plant's nutritional success. The pitcher is an insect trap and the acidic liquid contains enzymes including trypsin which break down the proteins of the trapped insect to a form that can be absorbed by the plant and utilized as a source of nitrogen and phosphate.

Each pitcher is somewhat tubular, swollen below and constricted above. During development the mouth of the pitcher is sealed by a lid which is eventually raised slightly above the pitcher's oblique mouth and serves both to

Below: A fly ventures on to one of the leaves of a Venus fly-trap. If it touches two of its trip-hairs the paired leaves will instantly close on it and crush it to death.

keep out the rain and to attract insects to the nectaries on its underside. The mouth has a hard glossy ribbed rim which ends in sharply-pointed teeth on the inside of the pitcher. Between these teeth are large nectar glands and below them, inside the pitcher, is a zone covered with flakes of wax. The trap is set! The insect is attracted to the plant by the promise of nectar and, reaching for the nectaries just below the rim, but unable to grip on the glossy surface, it falls into the digestive liquid below. It cannot scale the waxy pitcher sides or the downward directed teeth and so it dies and is digested. So effective is this trap that, in the course of evolution, these pitchers have become the sole habitat of several species of spiders, mosquitoes, gnats, small crustaceans, protozoans and algae, all of which have evolved to overcome the problems of living in such a potentially hostile habitat.

Carnivorous plants

Plants which live on peat bogs have nutrient deficiency problems similar to those faced by epiphytes, for they live on very wet areas dominated by species of bog moss *(Sphagnum)* and often raised above the level of the surrounding land like a giant sponge with the result that their only water supply comes from the rain and snow falling directly upon them. The few nutrients that are brought in in this way are rapidly taken up by the *Sphagnum* leaving little for the other inhabitants of the bog. In such a situation the carnivorous (or insectivorous) habit has again developed.

The peatlands of North America support ten or so species of the genus *Sarracenia*, characterized by pitchers which differ from those of *Nepenthes* in that all the leaves emerge from ground level in pitcher-like form and have conspicuously coloured upright lids. Their mode of operation is very similar to that of *Nepenthes* although it differs somewhat in *Darlingtonia*, a relative of *Sarracenia*, where insects are attracted by nectar to a downwardly-directed mouth and then fall into the trap because they try to escape through translucent 'windows' on the far wall of the pitcher.

North America also provides one of the most spectacular of the insectivorous plants, Venus' fly trap *(Dionaea muscipula)*. This plant has a modified leaf blade consisting of two lobes each fringed with stiff spikes. Each lobe has a marginal zone of nectaries whilst the rest of the surface is covered by deep-red digestive glands and three stout 'trigger bristles' per lobe. Attracted by the nectar, insects land on the leaf lobes. The triggers are not as simple as they seem, for this is the only plant that can count! If one trigger only is touched—as it might be by a raindrop, or a small insect not worth bothering with—nothing happens. But if one is touched twice in succession, or two of the three hairs are touched, the trap operates. Within a second the lobes have closed together and the marginal bristles interlocked and the insect is trapped, held close against the digestive glands which begin to secrete digestive enzymes.

Charles Darwin, who published a book on insectivorous plants in 1875, demonstrated the large quantity of digestive fluid produced by

Left: Many of the large South American bromeliad family are epiphytes, growing high up on trees to obtain good light. They often catch water in the centres of their rosettes.

Venus' fly trap by making a small hole at the base of a leaf lobe after an insect had been caught. He found that the digestive fluid was produced in sufficient quantity to flow down the leaf stalk for nine days.

Darwin also demonstrated, perhaps not unexpectedly, that the more food the common European sundew *(Drosera rotundifolia)* took in through its leaves the better it grew. This plant of nutrient-poor peatlands has leaves covered with 200 or so long, drumstick-like glands, usually called tentacles. At the tip of each tentacle is a drop of sticky secretion looking like a drop of dew (hence the generic name from the Greek, *drosos* meaning dew). Any insect landing on the leaf is caught on the sticky tentacles which bend over and envelop the unfortunate creature. It is then digested by enzymes secreted from the tentacles.

From the Northern Hemisphere and South America comes another group of insectivorous plants, the butterworts *(Pinguicula* species). These plants are to be found in a range of wet habitats and form small rosettes of sticky leaves. Insects become trapped when they land on the sticky butter-coloured substance on the leaf surface. They are 'glued' rather than held down while they are digested.

The genus *Utricularia*, the bladderworts, is related to *Pinguicula* but possesses a far more spectacular trapping device. Members of this genus can be found in aquatic and damp terrestrial habitats throughout much of the world, but particularly in the tropics. The bladders from which the plants take their name are scattered over the whole plant and are seldom more than a few millimetres long. *Utricularia vulgaris*, the greater bladderwort of Europe, Asia, North Africa and North America, is a widely studied aquatic member of the genus. In this plant the bladder, prior to trapping, encloses an empty flattened cavity apparently under a slight vacuum. At the mouth of the cavity is an inward-opening valve surrounded by sensitive hairs. If a small crustacean or insect larva happens to touch the sensitive hairs in passing, the valve immediately opens and the victim is sucked in, unable to escape because the valve closes after it. In time the larva dies and its body is broken down to be absorbed via star-like glands which line the bladder. The removal of water and minerals by these glands results in the bladder walls being sucked in so setting the trap again. Some tropical species of *Utricularia* make use of even temporary collections of water such as those which occur in certain flowers. They can be found straggling in a thread-like manner from flower to flower.

The use of animal traps by plants is not limited to the highly evolved flowering species. The soil fungus *Dactylella bembicodes* forms constricting rings, the cells of which swell as soon as an eelworm passes through. The worm is tightly held until the fungus invades and digests it. Other related species form sticky branches, to which the eelworms adhere, or spores, which stick to and germinate on the eelworm finally invading and digesting the animal.

The fungi as a whole are a group of plants that do not photosynthesize, so they must rely upon other organisms for food whether these be dead or living. They become parasites if the food is still living and saprophytes if it is already dead.

A little red club no more than a few centimetres high projecting through a mat of moss or grass is often worth careful excavation for below it may be the remains of a caterpillar, still recognizable as such but converted by the fungus *Cordyceps militaris* to 'solid fungus'. Another fungus parasitic on insects is *Aschersonia*. This has been used for the biological control of scale insects in Florida and elsewhere where the climate is humid enough to allow it to grow in a mat-like fashion over its host. Possibly the most familiar insect-parasitizing fungus is *Entomophthora muscae*. It is this fungus which causes the death of flies found adhering to windows that are seldom cleaned. A solid plug of fungus forms inside the fly and kills it. Shortly after the fly dies the fungus bursts through the body wall and large numbers of spores are released, forming a greyish 'halo' around the dead fly.

Saprophytes

Fungi are not the only plants to have given up photosynthesis. There is a liverwort, *Crypto-thallus mirabilis*, which completely lacks chlorophyll, except in its spores, and lives saprophytically with the aid of its associated (mycorrhizal) fungi in plant litter. And among the higher plants quite a number have given up photosynthesis for a parasitic or saprophytic way of life.

One family of saprophytic plants, which is found throughout the north temperate parts of the world and also extends as far as Malaya, is that which contains the yellow bird's-nest (*Monotropa hypopitys*). This plant has a yellowish waxy appearance and because it needs no light is found in the dense shade of coniferous and beech woodland. The below-ground parts are a nest-like mass of roots and associated mycorrhizal fungi.

Another plant, of similar common name and habit but unrelated, is the bird's-nest orchid (*Neottia nidus-avis*) of Europe and Asia. Again a plant not dependent upon light because of its saprophytic habit, it is to be found growing on humus-rich calcareous soils of shady beech woodlands.

Parasites

Total parasites, dependent for all their nutritive needs upon a living host, are widespread in the flowering plant kingdom. *Rafflesia*, a parasite on the roots of *Cissus* vines, is found in Malaya. It produces what is probably the biggest flower in the world, some 1m (3ft) in diameter and weighing around 6kg (13lb). A European relative of *Rafflesia* is *Cytinus hypocistis*, the yellow cytinus, which is parasitic on the rock or sun rose *(Cistus)*. It produces a globular head of bright yellow flowers below which are scales of yellow, orange or scarlet. The flowers

Top: The locked wings and hunched body of this dung-fly are typical symptoms of attack by the fungus **Empusa**, *which entirely takes over the fly's body. House-flies are frequently attacked.*

Right: Pitcher plants like this tropical **Nepenthes** *are slippery sided traps with incurving spikes which prevent insects from escaping; they fall into digestive fluid at the bottom.*

Far right: Once an insect has landed on a sundew leaf the sticky hairs hold it against any struggles. They then close in around it and acid secreted by the leaf digests the prey.

emerge just above the soil surface near the roots of the host.

There are about 100 species of the parasitic plant the dodder (*Cuscuta*) in the world, many of which are of considerable economic importance because of the damage they do to crops. *C. campestris* from North America intertwines amongst the stems of lucerne or alfalfa to such an extent that harvesting is difficult. Except for support immediately after germination dodder has no roots but winds around the stem of its host sending out haustoria or suckers into the stem in order to tap the nutrients passing along in the phloem. The flowers frequently smell of rotting meat to attract the flies which pollinate them.

The broomrapes are a large family of parasitic plants found throughout the Old World, with one American species, *Orobanche ludoviciana*, which was at one time relished as food by the

Left: Broom-rapes like Orobanche rapum-genistae are total parasites with underground roots attached to those of host plants; they only appear above ground when in flower.

Below: Lichens like Lobaria pulmonaria are remarkable examples of symbiosis, in which a fungus and an alga combine to form a distinct organism. Many grow on trees.

Pah Ute Indians. They are, fortunately, less important as crop pests than they used to be due to modern methods of cultivation, but some species which attack crops such as clover and peas are still often found in less-developed countries, and there are some tropical species which are still extremely destructive.

Toothworts (*Lathraea*) are parasitic plants of both Europe and Asia and were once thought by taxonomists to be closely related to the previous example. They are now, however, considered to be near relatives of the hemiparasites described below. Whatever their affinities, toothworts are fascinating plants. *L. clandestina* produces no aerial shoot and even some of its flowers may remain below ground level. *L. squamaria* has a stout erect flowering shoot with scale leaves bent over so as to enclose a chamber lined with water glands through which excess water is exuded. This exudation is necessary because the plant has no surface for transpiration and being a parasite predominantly on tree roots it needs to take in large volumes of water to be able to extract the food it requires. As waste water trickles down and evaporates the area around the plant often becomes white and encrusted with salts.

Hemiparasites

There is a large group of plants belonging to the family Scrophulariaceae which have not given

The final epitaph

Lichens are among the most successful symbiotic relationships between plants, the partners being an alga and a fungus which have become completely integrated. They have various characteristic forms, of which one of the commonest is the thin flat crust, like that growing on this gravestone. Crust-forming lichens typically grow on stone or rock, often on tiles; they are the earliest colonisers in very hard growing conditions like the Arctic, being tougher than either constituent part and capable of standing extreme cold, desiccation and heat. One remarkable example, the bushy **Ramalina** from the Negev desert, can tolerate over 80°C when dry, yet continue to photo-synthesise when part-frozen at −10°C.

The algal component of the lichen carries out photosynthesis to produce organic food materials; the fungus obtains mineral food from the material the lichen grows upon. When lichens grow on rocks the fungus eats into their surface. Eventually this leads to crumbling of the rock and this is the first step towards the creation of soil in which plants with normal roots can gain a foothold. Lichens grow very slowly on dry rocks and the larger examples in the photograph may well be a century old, although they are impossible to date exactly.

Many lichens like this are strongly coloured when growing in sunlight, often in shades of yellow and orange. These tints, which man has used for millennia as dyes, are due to a deposit of lichen acids in the cells nearest the surface. These act as filters to reduce the strength of light reaching the photosynthetic algal cells; the same lichen growing in shade will be green.

Crusty lichens obtain their water supplies from rain or run-off water. They are thus very susceptible to any impurities in the water, and in modern times dissolved sulphur dioxide, perhaps the commonest industrial effluent, has been found to damage and finally kill them depending on its concentration. Hence lichens are widely used as indicators of the degree of industrial pollution.

Above: The common mistletoe, Viscum album, is a partial parasite. It obtains moisture and food via suckers forced into the host tree, but carries out photosynthesis with normal leaves.

up photosynthesis and look quite normal plants above ground. Below ground, however, they rely upon their attachments to the roots of other plants for water and mineral salts. Such plants are usually called partial or hemiparasites and include such widespread groups as the eyebrights *(Euphrasia)*, red and yellow rattles *(Pedicularis* and *Rhinanthus)* and *Bartsia*. Mistletoe is another hemiparasite which has given up a root system and now attaches itself to the branches of its tree host, obtaining water and minerals from the xylem flow but producing its own sugars by photosynthesis. Mistletoes are very serious tree pests in many parts of the world.

Symbiosis

From a relationship where one partner gains at the expense of the other, as with a parasite and its host, it is but a small evolutionary step to a symbiotic relationship where two organisms live together to their mutual advantage. Such a relationship may be animal with animal, plant with animal or plant with plant.

Zoochlorellae (green algae, mainly freshwater) and zooxanthellae (yellow or brown algae, mainly marine) live in a symbiotic state with various aquatic invertebrate animals. From the relationship the alga gets some protection, exposure to the light and a source of nutrients from the waste products of the animal. But in some cases, as we shall see, it can lead to the plant's ultimate destruction. The zoochlorellae in the tissues of *Hydra* is a symbiotic relationship that is well-known at the most elementary biological level. The symbiotic relationship in the marine flatworm *Convoluta roscoffensis* is less well-known. This flatworm, although only a few millimetres long, occurs in such numbers along the beaches of Brittany in France that it gives the sand a greenish tint. This is not due to the worm's natural colour but to all the algae which the worm has eaten and which have then invaded its tissues. At maturity the worm is no longer able to feed so it begins to digest its accumulated algae, eventually using them up quicker than they can reproduce. The result is that soon after laying its eggs the worm dies having used up its algal food reserve.

Other animals with an algal symbiont include some corals, the giant clam *(Tridacna derasa)*, tropical jellyfish such as *Cassiopeia* and some sea-slugs such as *Aeolidiella glauca*.

Plant with plant symbiosis has already been mentioned in referring to the mycorrhizal fungi which live in association with the roots of higher plants such as orchids and heathers.

Anthoceros is a liverwort which has a symbiotic relationship with the blue–green alga *Nostoc*. It seems probable that the alga fixes nitrogen from the air for its partner but other

than protection it is not clear what advantage *Nostoc* derives from the relationship. In a similar way bacteria of the genus *Rhizobium* fix nitrogen in the roots of leguminous plants such as peas, beans and clover and probably obtain carbohydrates in return.

The bacteria form small nodules on the roots which are clearly visible if dug up. Other bacterial root nodules occur in the madder family *(Rubiaceae)*, where an unexpected phenomenon occurs—without the bacteria, the plants become dwarfed. In this case, what we call "normal" growth is entirely dependent upon the symbiotic bacteria.

Symbiosis with fungi is particularly important to coniferous trees, where the association of fungus strands and tree roots creates a knobbly, sometimes coral-like growth known as mycorrhiza. The fungus forms a sheath of tissue around the tree roots and also penetrates between the cells of their outer layers. A large variety of "toadstool" type fungi associate with trees in this way. This symbiosis is especially important to the trees early in their life: seedlings in fungus-free soil are stunted and grow very slowly. The fungus stimulates root penetration and enables the roots to absorb more food from the soil, while receiving carbohydrates from the tree in return.

Lichens

One of the most successful symbiotic relationships is that found in lichens, those predominant colonizers of tree trunks and bare rocks. Here the symbionts are an alga and a fungus, the former providing the photosynthetic products and the latter the inorganic nutrients required for the maintenance of the lichen. Lichens are an interesting taxonomic problem because two separate organisms are being referred to by one scientific name. In spite of their compound origin the shrubby reindeer lichen *(Cladonia)* will always reproduce to form another *Cladonia* plant and the orange alga *Xanthoria* to form another *Xanthoria* and so on.

The body of a lichen is called the thallus and a thin section through this in a fairly typical species such as the grey dog lichen *(Peltigera canina)*, which is found on grassy banks, will show an upper and lower area of fungal hyphae with a layer of green alga sandwiched just below the upper surface.

Although a lichen may get some nutrient from the substrate on which it is living most of its inorganic requirements are obtained from rainfall and run-off water flowing over it. Lichens are, therefore, opportunists grabbing all they can from any water that flows over them. This has been their downfall. As the world becomes more polluted and the air carries more gases such as sulphur dioxide and the oxides of nitrogen, all of which form acids when dissolved in water, lichens are being poisoned out of existence and city centres and industrial areas are now 'lichen deserts' with few if any species to be found.

Lichens reproduce in two ways, asexually and sexually. Asexual reproduction involves the production of powdery soredia or slightly larger isidia which contain both algal and fungal elements. Sexual reproduction is by fungal spores alone and these will only form a new lichen if the normal algal partner happens to be present where they germinate.

Their ability to live on bare rock means that lichens are often the first colonizers of newly exposed areas, providing by their eventual decay the humus for new soil—the home for more conventional plants.

LEGUMES
pods for protein

Legumes have been important constituents of human diets for many thousands of years. Neolithic man in the Near East was certainly eating pods of wild kinds such as vetchlings and shauk at least 10,000 years ago. The predecessors of the Incas of Peru were eating kidney beans some 7,000 years ago. Broad beans have been eaten in Europe since the Iron Age—the ancient Greeks and Romans, however, believed that eating them would dull their intelligence, and Pythagoras regarded the eating of beans as a sign of moral weakness! The inclusion of leguminous plants in crop rotation over the past few hundred years was one of the great advances in agriculture, and today they represent a vital source of protein for an undernourished world.

Botanically the legumes belong to the order Fabales, an enormous group of some 700 genera and about 14,000 species (almost 5 percent of all the known flowering plants). This family is divided into three subfamilies. The Papilionaceae (or Leguminosae) have flowers which are mostly bilaterally symmetrical, the petals often being unequal in size with a hood and a keel. In the other subfamilies, Mimosaceae and Caesalpinaceae, the flowers are quite different; but almost all produce the characteristic pods enclosing the seeds, though these may range from soft to woody, and from straight to spiral. They may be fleshy, as in runner beans, or relatively thin and fast to dry out, as in the gorse or furze, *Ulex*. Most species have divided leaves, sometimes into three rounded leaflets, as in clover, more often pinnately with a number of leaflets on each side of a mid-rib, as in vetches and acacias. Some genera, like gorse and many acacias, have well-developed spines which originate as modified leaves. In temperate regions most legumes are herbaceous, but in the tropics the greater proportion are trees.

Old foods and new foods

The garden pea, *Pisum sativum,* is known from Stone Age caves in central Europe and from Ancient Troy. The various species of bean, *Phaseolus,* are eaten in many parts of the world, as is the lentil, *Lens esculenta,* which is thought to have a western Asian origin. There are many 'pulses', legumes grown for the mature seeds which are suited to drying and storage as a food source, for example butter and Lima beans, black and green grain or mung beans. The origin of the peanut, *Arachis hypogaea,* is not clear but it may have originated in South America, where a number of related species occur naturally, or in south China or Africa. The soya bean, or soybean, *Glycine max,* originated in Northeast Asia but, like many major crops, is now mainly grown away from its area of origin, today very extensively in the USA.

Legumes are important foods partly because their seeds often have a relatively high protein content compared with other crop plants. It is, however, not only the quantity of protein in the diet that is important, but also its quality; that is to say the proportions of the different amino-acids that make up the protein. Man requires over twenty different amino-acids to build the proteins he needs, but eight of them—the essential amino-acids—he cannot make for himself and must obtain from the protein in his food. All the essential amino-acids are plentiful in most meats and meat products such as milk and eggs, but proteins from plant seeds usually have four of them present in small amounts only—threonine, lysine, tryptophan and methionine. A diet based on plant seeds would thus require a large amount of protein in order to obtain sufficient of these amino-acids. On the other hand as a major supplement to animal proteins, seed proteins are of great benefit.

When a cow feeds on plants some of the plant proteins are used to build more cow, ie meat, but much will be lost by excretion. If the cow is eventually eaten for meat only a relatively small proportion of the original plant protein will be available, perhaps about 20 percent. If, however, man was to eat the plant directly all the protein would be potentially available, although in practice not all would be in an assimilable form. Thus, where there is a shortage of protein for human consumption eating the plant material directly is a much more efficient use of a scarce resource.

One legume which is being increasingly used as a protein source is the soya bean. The protein is extracted from the beans, con-

Myroxylon balsamum var. pereirae
Balsam of Peru

Right: The winged or asparagus pea,
Lotus tetragonolobus, *is one of the legumes with fleshy pods which are eaten by humans. It has been cultivated since the 16th century.*

Above: Many legumes exhibit a remarkable example of symbiosis with a soil bacterium which forms nodules on the roots and is capable of 'fixing' atmospheric nitrogen.

Left: Brooms and their relations, like this New Zealand species, have hard woody pods which disperse the seeds explosively. Many are ornamental plants grown in gardens.

centrated and processed, and is then used as a substitute for meat, which it may, upon suitable preparation and cooking, be made to resemble; steak which has never seen a cow! It should, of course, not be forgotten that soya beans may be cooked and eaten directly, eliminating the need for industrial processing, providing that one does not need the illusion that it is meat that is being eaten. It is not only seeds that may be treated in this way. An important potential protein source comes from the leaves and seed pods of many plants, including legumes, which are at present discarded when the seeds have been separated. Protein may be extracted from these just as from soya bean.

Legumes and nitrogen fixation

Legumes are also important for their ability to metabolize nitrogen. They have the property of producing combined nitrogen compounds from atmospheric nitrogen, N_2—a process known as nitrogen fixation. Nitrogen is an essential component of the amino-acids which make up proteins, the enzymes and structural elements essential for the living cell. With the exception of the nitrogen-fixers, organisms can only take up nitrogen in some combined form: plants mainly as nitrate, NO_3, or sometimes as ammonium compounds, NH_4, taken up by the roots; animals as ammonium compounds such as amino-acids and proteins from plants or from other animals.

Some combined nitrogen may enter the soil from the weathering of rocks containing nitrates and some may be formed in the soil by lightning, reaching the soil dissolved in rain, but these two sources are small compared to the amounts formed by biological nitrogen fixation.

Nitrogen fixation in legumes is the product of a symbiotic relationship between the plant roots and the bacterium *Rhizobium*. The bacteria occur freely in the soil, and may be attracted to the roots by a growth substance produced by the plant. The bacteria enter the root by penetrating the root hairs, and migrate into the cortex of the root. A growth substance secreted by the *Rhizobium* causes the root cells to divide and grow to form a root nodule. Nodules are usually roundish structures, sometimes branched, normally situated on the fine lateral roots. Vascular strands from the plant root run into the nodule, in the centre of which are the large number of cells containing *Rhizobium*.

The biochemical mechanism of nitrogen fixation is complex and as yet not fully understood, but its net result is the splitting of molecular nitrogen and its incorporation into amino-acids. The plant thus obtains a supply of amino-acids and the bacterium receives organic compounds from the plant as energy sources. If an active root nodule is cut it will appear to 'bleed'—the cut interior of the nodule being in places clearly red coloured. The substance concerned is none other than a haemoglobin, the compound responsible for the transport of oxygen in our blood. What its function is in a plant root is not certain, but it may be involved in controlling the oxygen concentration within the root nodule.

The nitrogen so 'fixed' will initially all be within the plant. It will only become available to the other parts of the eco-system either when the plant is eaten by an animal or when the plant dies and its breakdown releases nitrogen compounds into the soil. This input of combined nitrogen will be especially important where legumes are growing in nitrogen-deficient soils, such as those of recently formed sand-dunes. Legumes are also important where the harvest of agricultural crops results in the loss of combined nitrogen from the soil. This is the reason for including legumes in many crop rotation schemes. The amounts of nitrogen added to the soil may be considerable, often as much as 100–200kg per hectare (88–176lb per acre). The legumes, such as alfalfa, *Medicago sativa,* and clover, *Trifolium,* are often grown as a winter crop and then ploughed into the soil, or are grown in a mixture with grasses so that some combined nitrogen is also passed to grazing animals.

It should be remembered that legumes are not the only organisms that can fix nitrogen; there are a few non-legume higher plants that can do so, as can a number of micro-organisms, mainly bacteria and blue–green algae.

Among the dreams of plant scientists is the possibility of producing crop plants such as wheat which are receptive to nitrogen-fixing bacteria; something which might be done by fusing or otherwise combining the cells or nuclei of, say, a wheat plant and a bean. Such nitrogen-fixing cereals would not only vastly improve soils they were grown in but immensely reduce the need for fertilizer application, notably of nitrates which have created so many side problems by their excess amounts draining into watercourses and lakes. That would be a revolution indeed.

Phaseolus vulgaris
Kidney Bean

ROOT VEGETABLES
the country stores

When early man was evolving as a hunter and food gatherer, he discovered that certain plants store their food reserves in swollen roots and similar underground structures. The advantage to the plant is that it can survive in this way through either a very cold or a very dry season of the year, and resume growth when the weather becomes warmer or wetter. The human food-gatherers found that if they dug up these roots they had a handy source of nourishment to see them through a bad time of year, and one that was easily stored. As farming developed, root vegetables were cultivated as high-yielding easily-raised crops throughout the world and settlers naturally carried them to new countries. Many are now grown thousands of miles from their first homelands.

When is a root not a root?

Only a minority of root vegetables are true roots in the botanical sense. Carrots, turnips, beetroots and radishes are in fact swollen portions of their plant's underground root systems. But onions are bulbs, that is groups of colourless leaves modified for storage of food reserves and the plant's true root can be seen growing out beneath them. Potatoes, yams and tapioca or cassava are all tubers, that is swollen portions of underground stems. They have the capacity to send out both roots and shoots after their resting period, so such tubers can be detached and planted in order to establish fresh crops. In contrast, true roots and bulbs are usually raised afresh each year from seed.

Most root vegetables store the bulk of their nutrients as insoluble starch, a carbohydrate that is readily digested, after conversion into soluble sugar, by man. A few, such as beetroot, have a variable proportion of sugar, and hence are sweet to the taste. Some, such as tapioca, are actually poisonous if eaten raw, and must be cooked to make them harmless and wholesome. Root vegetables as a whole are a valuable source of essential vitamins, especially vitamin C. They also contribute mineral elements to our diets, but are poor in fats and proteins.

Biennial root crops: radishes, carrots, turnips and beetroots

Many root vegetables have a two-yearly or at least a two-seasonal life cycle, and hence are called biennials. During its first summer the seedling grows, but does not flower. Foodstuffs that might have supported flowers and fruits are diverted instead to the roots, which swell out as they become storage organs. Next spring the plant is able to draw on this reserve, so it can open flowers and ripen seeds profusely early in the season. Therefore its seedlings get a good start in competition with other plants and in their struggle against a difficult environment. After seeding, the parent plant dies.

Man intervenes in this simple life cycle by harvesting the swollen roots for his own nourishment, but at the same time he is careful to keep a breeding stock of plants that he allows to mature and bear seed for his next crop.

The familiar red-rooted radish, *Raphanus sativus*, which matures in a matter of weeks, but will quickly 'run to seed' unless uprooted promptly, is a good, if rather high-speed, example of this kind of husbandry. Radishes come originally from the Mediterranean region, where their pattern of short fast growth-periods, followed by pauses of rest, is adapted to short spells of rain under a hot climate.

Carrots are grown on a similar plan; they can,

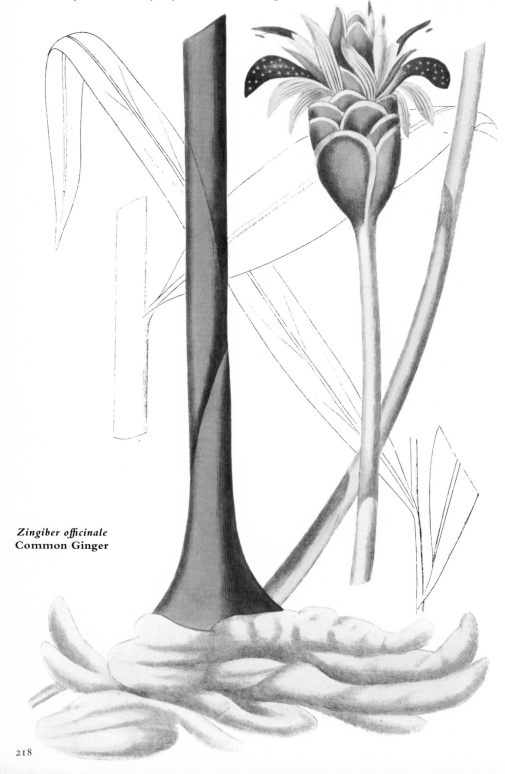

Zingiber officinale
Common Ginger

if desired, be stored for several months after harvesting, for their corky skin resists water loss. The wild carrot, *Daucus carota*, grows on dry chalky soils in southern England, as a feathery-foliaged white-flowered herb. Its thin whitish roots, developed to resist seasonal droughts, bear little flesh, but have an unmistakable carrot flavour.

Turnips, *Brassica rapa*, and their near-allies the yellow-fleshed swedes, have been grown for centuries as tasty root vegetables, especially for winter use. They are possibly the roots first used by man, for their remains have been found on Neolithic sites. In the eighteenth century a reforming landowner, Viscount Townshend, who inevitably gained the nickname 'Turnip Townshend', discovered that they could be grown on a grand scale for feeding raw to sheep and cattle. This enabled farmers to maintain larger stocks through the winter, when grass was scarce. These roots augmented the small supplies of hay, hitherto the only easily-stored winter fodder, and made farms far more productive and profitable. As a rule, turnips are harvested in autumn and stored in clamps which are long stacks covered first with straw, then with a layer of earth, to keep out the frost. In Scotland, sheep are allowed into the fields to eat these roots where they grow.

Wild beetroots, *Beta maritima*, can be found on European seashores growing as straggly

*Below: The turnip, **Brassica rapa**, has been eaten since ancient times; it was developed from one of the wild cabbages of northern Europe. Though filling, its food value is low.*

Above: The potato tuber is a storage organ from which the plant can reproduce readily: shoots from it grow into new stems, and man's crops are always grown from sprouted potatoes.

Left: The potato, Solanum tuberosum *(related to tomato and nightshade), originated in the uplands of South America. It can yield greater food value per acre than any cereal.*

grey–green weeds that resist blazing sunshine on a dry, salty and sandy soil. Under cultivation, growers have developed three strains, all with much larger roots than their wild ancestors. Red round beetroots make tasty sweet table vegetables after cooking. Sugar beets, which are white and tapered in shape, like carrots, are a main source of refined sugar. After this has been washed out at the factory, their spent roots are used as cattle fodder. Spinach beets are grown for the tender leaves that spring from their roots, and are cooked for use as green vegetables, like spinach.

Onions

The common onion, *Allium cepa*, originated in the hot dry climate of the Near East, and its fleshy bulb stores water, as well as food, through prolonged droughts. This explains the presence of several layers of thin waterproof skins surrounding the white modified leaves that compose the bulb. Onions are raised from seed, sown in spring and harvested in autumn as an easily-stored source of both food and flavouring. Those left for seed open ball-shaped clusters of pretty bluish-white flowers on tall stalks, followed by papery white pods holding many small black seeds.

Potatoes

The value of the potato, *Solanum tuberosum*, as a staple food for man was discovered in prehistoric times, apparently by Inca Indians who lived in high-altitude regions on the slopes of the Andes of Peru, in subtropical latitudes but above the hot levels where the main tropical tubers are cultivated. Potatoes were carried wherever the Incas' pattern of primitive farming spread. They were introduced to Europe by the first Spanish explorers, and are now grown all over the world. Scores of varieties have been bred, but none is frostproof, and in all cold countries they must be stored away from winter cold, usually in clamps covered with straw and earth.

Potatoes bear pretty white or blue-tinged flowers, with reflexed petals, a prominent central cone of yellow anthers and a green pistil. They ripen many seeds in round green berries shaped like little tomatoes (which are in fact poisonous). But in practice they are only raised from seed by breeders seeking to establish new varieties, because seedlings start growth slowly and vary in character. The main potato crop is raised by cultivating the land, then burying small tubers, which are called 'seed potatoes' (though they are not seeds at all), just below the surface of the soil. Shoots spring up from little buds, called 'eyes', on the surface of the tuber, and roots break out at the same places. A dense low bushy plant results. This sends out, just below ground level, odd white side stalks, on each of which a new tuber develops, close to the soil surface. About midsummer the farmer 'earths-up' his crop, using a special plough to pile soil over these new tubers, which would otherwise become hard, greenish-brown, badly-flavoured and slightly poisonous; then in autumn, when the foliage has withered, he uses another plough, or machine, to raise his heavy underground crop. Similar practices are followed in gardens.

Tropical roots

Tapioca is obtained from the cassava or manioc plant, *Manihot esculenta*, native to Central and South America, but now grown in every tropical country, especially West Africa and Southeast Asia. It is planted as stem cuttings, on well-cultivated and fertilized land, and takes 12–15 months to mature. Its jointed shoots quickly grow tall, often exceeding 2m (6.5ft), and the crop looks like a forest of waving stems bearing large compound leaves. Clusters of long sausage-shaped roots develop below ground, giving a heavy yield of over 30 tonnes per hectare. These starchy roots hold small amounts of poisonous prussic acid and cannot be eaten raw. American Indians prepare them for food in two ways, either slicing and boiling them to wash out the poison, or drying them and then pounding them into a coarse meal called manioc. The familiar tapioca used for puddings is prepared for storage, or export from tropical countries, by an elaborate factory process of rasping the roots, washing out the starch grains, then drying them to form a meal. This is next rocked in hammocks, causing the grains to form round globes, which are cooked gently to make them hold together. The waste is fed to pigs.

Yams, which are a staple food in tropical Africa and the West Indies, are the large tubers of a climbing plant called *Dioscorea batatas*, originally native to China, and its allied species. Yams are planted as tubers which produce slender stems that are trained up stakes. After about 10-months growth their blunt-ended tubers (often 40cm (16in) long by 10cm (4in) thick), mature and are dug up. The yams are stored in racks, with air circulating round them; as long as they are kept dry they will not sprout again. They have a food value comparable to that of potatoes, holding about four-fifths by weight of water and one-fifth nutritious starch, with a little protein.

Taro, *Colocasia esculenta*, the staple root crop of the Pacific Islands, is also grown in the West Indies under the names of eddoes and cocoes. It is a monocotyledon related to the arum lily, and grows as a tuft of tall stalks bearing huge heart-shaped leaves. Its oval roots yield the starch that is made, in Polynesia, into a thin porridge called 'poi'.

A number of other tropical plants produce edible roots, including the sweet potato (a morning glory relation); uca (a wood sorrel relation); ulluco (of the family Basellaceae); and arrowroot (related to the marantas, popular as houseplants).

WEEDS
pioneers and colonisers

The term weed is rather imprecise because it means different things to different people. To the farmer weeds are the unwanted wild plants growing in his crops, while to the gardener all the plants he is not cultivating are weeds. However plants cultivated as crops or as garden flowers may persist spread, and themselves become weeds in certain situations. To the uninitiated perhaps all wild plants are weeds because they are apparently unproductive and of no particular use to man.

The best definition of a weed is a plant which is growing where it is not wanted. In general weeds are colonizing plants that thrive in the disturbance created by man's activities. They are successful because they exploit man's activities both for their habitats and their dispersal, and in many cases because they have evolved with man in the old centres of civilization.

Weeds as colonists

Weeds are usually associated with an aggressive ability to spread quickly and become a pest and may be difficult to control or eradicate. They do actual harm to cultivated plants by shading them out, depriving them of sunlight and by competing with them for nutrients and water. Each climatic zone of the world and each particular soil type has its own particular prevalent weeds which are able to exploit these conditions most effectively. The most aggressive weeds in a country are very often not natives but are introduced aliens. This is perhaps because the normal controls such as herbivorous insects have been left behind in the country of origin. Many weeds are virtually cosmopolitan, having been distributed by man's movements round the world.

Weed species have a number of common characteristics. They are usually colonists of open or disturbed ground so they are particularly noticeable on cultivated arable land or in gardens, or in other situations where the natural vegetation cover is broken up and soil exposed. These conditions occur temporarily on building and construction sites, on waste ground and roadsides and on rubbish tips. A relatively small number of weeds can compete

Right: One necessary ability of a successful weed is to grow almost anywhere its seeds may land. The valerian (Centranthus ruber) is fond of crevices in stone or rock.

Below: As well as having powers of regeneration from damaged roots, the dandelion produces great numbers of airborne seeds with which it colonizes virgin ground, often at a great distance.

Taraxacum officinale
Dandelion

in closed turf such as lawns. Ponds, lakes and rivers also have their own weeds which may become very troublesome in navigational waterways.

Dispersal of weeds

The capacity to produce prolific quantities of seed is a quality which enables weeds to rapidly colonize open situations. This may be helped by the production of several generations of fruiting plants during a year. Perhaps the most important factor that contributes to the success of a weed is the ability to flourish and reproduce in a wide variety of different conditions. Wind dispersal of seeds enables rapid spread and colonization of new habitats. A number of successful species have evolved seeds that can be carried by the wind, especially those which have a parachute of hairs such as the seeds produced by thistles, willowherbs and fleabanes.

Many seeds that are eaten by livestock or birds may remain viable after passing through the digestive system of these animals and so dispersed. In some cases the viability of the seeds is actually enhanced by this process. Animals are responsible for distributing other weeds by carrying the seeds externally. Seeds or fruits that bear hooks or barbs are designed to become entangled in the coat of animals. Small seeds can be picked up in the mud that adheres to animals' feet or to the wheels of man's vehicles.

Vegetative reproduction may be a significant means of multiplication of weeds especially where conditions are relatively uniform. Some of the most ineradicable weeds can regenerate from fragments of rootstock as in the creeping thistle *Cirsium arvense*. Ploughing and harrowing breaks up the roots and spreads the plants by scattering the fragments. Similarly the weeds of the genus *Oxalis* produce tiny bulbs or 'bulbils' on their stems which readily break off to produce new plants.

Man is undoubtedly the most significant factor in the dispersal of weeds both within a country and round the world. It is man's own commerce that has introduced alien plants into new countries. Foreign plants may be introduced into a country as cultivated plants which are deliberately imported or as foreign weeds which are accidentally introduced. Cultivated plants include those imported as garden plants, as crops or as fruits or seeds for processing. Weeds may be introduced with almost any product that is capable of harbouring the seeds. This is especially true of seeds or grain for cultivation or animal foodstuffs, which may include impurities. Weed seeds may also be attached to wool, hides or furs; included in packing materials; or adhering to people, their vehicles or livestock. In this way a number of weeds have become so widespread that it is now difficult to determine where they originated. Examples of species which are more or less cosmopolitan are bracken, *Pteridium*

Above: Besides enjoying the ground disturbed by cultivation, annual weeds are ready colonisers of bare ground like that of rubbish tips and building sites.

aquilinum, knotgrass or ironweed, *Polygonum aviculare*, and annual meadow grass, *Poa annua*. The widespread weeds of the temperate regions of the world, such as field convolvulus, *C. arvensis*, may also grow in mountain areas in the tropics.

When Europeans first emigrated to North America they took with them their weeds of cultivation so that many European cornfield weeds, such as couch grass, *Agropyron repens*, and groundsel, *Senecio vulgaris*, are now troublesome weeds in USA and Canada. Similarly the weeds of New Zealand are largely European in origin with species such as ragwort, *Senecio jacobaea*, and spear thistle, *Cirsium vulgare*, being a particular nuisance. In Africa, where the climate is rather hostile, European plants such

Shoddy aliens

One way in which seeds are transported is on animal coats, and man spreads them even further by importing wool. Waste from wool factories is made into shoddy, then used as a slow-acting nitrogenous fertiliser; and land manured with shoddy may grow unfamiliar plants like the South American Verbena bonariensis *here flowering among young trees on an apple farm in England.*

as the ribwort plantain, *Plantago lanceolata*, may occur as weeds in wet places.

Weeds of husbandry

A good illustration of the way in which weeds have been distributed around the world by man is given by the plants which adhere to sheep's wool. Commercially-useful wool is produced by merino sheep which were first raised in Europe in Spain. During the eighteenth century

the king of Spain made gifts of merino sheep to the Dutch governor of South Africa and to George III of England. Sheep were then taken from England to Australia in the eighteenth century and were later crossed with rams from the Cape. The Spanish took sheep to South America, from whence they were taken into the USA.

During all these movements of animals the seeds of weeds were transported on the sheep's fleeces. The special adaptations of these weeds include hooked or spiny fruits, spiny flower-heads or tiny seeds that adhere to the greasy wool. In this way many weeds of Mediterranean origin were distributed far and wide.

In Australia, before European settlement, the impact of man and animals on the native vegetation was slight. However when the Europeans arrived with their sheep, cattle and rabbits and their pastoral way of life this native vegetation was rapidly modified and replaced by the alien weeds carried by the sheep. These Mediterranean aliens were very successful pioneer species that were already pre-adapted to exploiting man-modified environments in the old centres of civilization in Europe. The rapid increase in sheep numbers during the second half of the nineteenth century distributed the alien weeds over a large area. The use of fertilizers such as superphosphates also benefitted the leguminous weeds which could more fully exploit the nutrient rich conditions. One of these weeds is subterranean clover, *Trifolium subterraneum,* a Mediterranean annual, which was first reported in Victoria in 1887 and is now widespread in Australia, extending far beyond the limits of the Mediterranean-type climate. More than 50 different strains of this plant have now been identified in Australia.

The weeds which are now so abundant on the sheep pastures of the main wool-producing countries of the world, ie Australia, New Zealand, South Africa and South America, are further distributed by the export of wool. The fleeces are imported into Europe and elsewhere for the production of woollen goods. On arrival at the woollen mills the fleeces are sorted and treated to remove grease, grit and dirt and also plant seeds. In spite of rough treatment many seeds remain viable after the cleaning process. At each stage in the handling of the wool there are opportunities for seed to fall in places where it can grow, for example at the port of entry or on the waste tips of the woollen mills. However the most productive source of weeds is the wool waste. Grey shoddy is the refuse of the wool combing sheds and it is used extensively in Europe as a manure especially on light sandy soils where it retains moisture and breaks down slowly. This wool waste contains a large proportion of plant material and viable seeds which may germinate in the fields as alien weeds. The variety of weeds introduced into Europe in this way is enormous and includes many very spectacular weeds in addition to the ubiquitous weeds of Mediterranean origin. Farmers who use shoddy in the market gardening or hop-growing areas of southeast England are familiar with some of the more extraordinary 'shoddy aliens' such as the South American barbed wire plant, *Xanthium spinosum* or the deadly poisonous thorn apple, *Datura stramonium.*

Weeds of arable farming

The weeds of cornfields and arable land form another interesting category because many of these plants have had a long association with man. They are generally dependent on the disturbance created by man when he plants his crops. Some of the abundant arable weeds found in Europe were themselves once cultivated species in Neolithic or Bronze Age times for example: the rye brome grass, *Bromus secalinus,* which was grown as a cereal; fat hen, *Chenopodium album,* which was grown as poultry food and the false flax or gold of pleasure, *Camelina sativa.*

Many of the other cornfield weeds were difficult to eradicate in the past by seed screening because their seeds so closely resembled wheat in size and shape. In some cases it was highly desirable to eliminate them from the crop because of their poisonous seeds. The pink flowered corncockle, *Agrostemma githago,* contains a poisonous glucoside in the seeds which spoilt the flour. The field cow-wheat, *Melampyrum arvense,* which has seeds like blackish wheat grains, used to render the flour discoloured in some districts of southern England. The prevalence of such weeds was due largely to primitive agricultural methods and where modern methods of seed screening and weed control are employed there are relatively few troublesome weeds.

Weed control

Chemical control of weeds by herbicides is largely practised in gardens or on arable land. The great advantage of herbicides is that they are labour saving, but there are dangers inherent in their use—resistant strains of weeds will develop or susceptible weeds may be killed and eradicated but then replaced by resistant weeds that become rampant when freed from competition.

Weed control may be non-selective, for example by using a flame gun or by using a drastic herbicide such as sodium chlorate. These methods are usually employed in situations where vegetation is not desired at all such as paths, car parks, railways and other artificial places. Selective weed control by herbicides is more appropriate in crops and gardens. Contact herbicides act directly on the plant by killing only those leaves contacted. The second basic type of chemical is translocated herbicide which acts more slowly and which is absorbed by the plant and moves within the plant before taking effect. Some modern chemicals such as diquat and paraquat exhibit characteristics of both types of herbicide. Treatment with these chemicals may be direct on the weed foliage or on to the soil surface.

An early chemical herbicide was copper sulphate which was employed in the early twentieth century to control charlock in cornfields. In the 1940s it was discovered that certain synthetic growth-promoting chemicals, phenoxyacetic acids, would act as herbicides and were selective in action. Factors that affect the amount of herbicide that is retained on a leaf include the leaf shape and position, and the nature of the leaf surface, i.e. whether it is waxy or hairy. Such differences can be exploited to make chemicals selective in their action. There is now a great variety of modern herbicides in use to protect crops employing many different mechanisms to ensure that the chemicals are selective weed killers.

It is unlikely that man will ever be entirely free of weeds. The inherent adaptability of weed species ensures that they will exploit every situation. In his efforts to grow productive crops man creates vast areas suitable for colonization by weeds. However in his efforts to eliminate weeds he progressively favours the most resistant and ineradicable ones.

*Right: The plantain, **Plantago major,** spreads very readily by seeds and, like most successful weeds, will grow in all sorts of soils and locations—even in the cracks between paving stones.*

GARDEN FLOWERS
plants for pleasure

From the earliest times man has appreciated the beauty of flowers. Their attraction arises from their marvellous symmetry and variety of form, their range and intensity of colour and their pleasing perfume too. Their transient nature—springing suddenly to perfection, then fading—enhances their charm, as expressed by the Scottish poet Robert Burns:

But pleasures are like poppies spread—
You seize the flower, its bloom is shed.

Historical background

Originally all flowers were plucked from wild plants or shrubs, but at surprisingly early periods, in various lands, people began to cultivate flowering plants within the protection of their gardens. The first gardens were fenced or walled enclosures devoted to plants that met other needs—fruits, roots and vegetables to eat, spices for flavouring food and medicinal herbs to cure ailments. Gradually the purely decorative plants claimed a small space, which tended to spread until, today, for millions of town and suburban dwellers, a garden means a flower garden and nothing more.

Flowers are perishable, and so their early history under cultivation is only known from sculpture, pottery, paintings or literature. The frequent appearance of flowers as art forms in carvings on ancient Egyptian and Assyrian temples proves that their cultivation was understood as early as 3000 BC, even though this meant reserving precious irrigated land from the production of essential food. Until recent times flower growing has always been associated with a wealthy ruling class or religious foundation which alone could afford to set aside and tend land for pleasure rather than economic production. Hence the links between attractive flowers and palace, monastery and temple gardens, assuring their care and propagation down the centuries.

The history of garden flowers reflects the story of mankind. In the western world the Greeks and then the Romans, inherited both the plants and the methods of tending them, that had been developed in Egypt, Syria and Mesopotamia. These skills were passed on to the monks who tended flowers in their cloisters through the Dark Ages. On the other side of the world, Chinese and Japanese gardeners were developing their own peculiar yet charming gardening patterns and traditions, based on plants native to eastern Asia. In Central and South America the American Indians, equally isolated from both these movements, were cultivating coloured varieties of dahlias as well as the maize and potatoes they needed for sustenance.

The great age of exploration, from AD 1450 to 1800, changed all this. Navigators who discovered new lands brought back seeds, bulbs and living plants to enrich their homelands. Though many introductions failed, others throve exceedingly. Garden flowers today are an astonishingly cosmopolitan collection, many found in every country, regardless of its climate. Leading flowering plants include chrysanthemums from Japan; pelargoniums and gladioli from South Africa; dahlias from South America; hyacinths from Asia Minor; and roses originating from European, Asian or North American wild stocks. Interbreeding between related species on remote continents has added both variety and vigour, and trade in new kinds is truly international.

All garden flowers originate in the wilds and, given sufficient care, any wild plant can be grown in a garden or greenhouse. But in practice only a small selection of wild flowers feature in everyday gardening. These have been chosen, by trial and error, as good material for

Rosa canina
Dog Rose

Right: Modern roses like Ena Harkness have an extremely complex ancestry mainly due to breeding from different species; but double roses were first cultivated from wild 'accidents' produced by species like dog rose, by the Romans.

both the nurserymen who raise them and the gardeners who plant and enjoy them. Though some keen gardeners tend their flowers through all stages of their life history, most prefer to buy seeds, bulbs or well-rooted plants from professional horticulturists. The specialist looks after all the difficult stages and copes with their problems, while the ordinary gardener enjoys the final, relatively easy stages of care and admires the result. Many popular flowers are only half-hardy. In this case the nurseryman cares for them during the winter season when frost can kill tubers or other breeding stocks and the outdoor gardener takes them over for the warm frost-free summer when they blossom.

An acceptable garden plant, as well as being attractive, must be easy to raise in bulk and should survive either seedling or transplanting stages easily. It must remain vigorous under varied conditions of soil and climate and be free from pests and diseases. The same basic plant should be available in a wide range of shapes, sizes and colours, for everyone enjoys variety. Most of the flowering plants grown in gardens today are the outcome of ruthless selection aided by planned breeding to obtain new strains for still more exacting choice.

Woody perennials

An important group of garden flowers grows as shrubs or woody plants with perennial stems surviving for many years. Roses and rhododendrons predominate in lands with temperate climates, but cannot stand tropic heat. In warmer countries hibiscus is grown instead. Woody-stemmed plants have one great advantage—they can be increased by grafting choice varieties on to common stocks. Once a desirable new kind has been discovered or artificially bred, it can be quickly increased and distributed to every suitable part of the world. Such a variety, increased by vegetative means, is called a clone or cultivar.

This is well illustrated by the handsome strains of double roses known as Bourbons. Late in the eighteenth century French farmers on the Indian Ocean island of Bourbon (now called Réunion) developed the pleasing practice of protecting their fields with hedges of spiny roses. No rose was native to the island, so they imported a Chinese rose, *Rosa sinensis* (now known as 'Parsons pink perpetual'), from its far eastern homeland. This was carried in by European trading ships on their homeward voyages. About the same time Arab traders brought from the Persian Gulf another choice strain, *Rosa damascena bifera*, nowadays called 'Pink autumn damask' or 'Four seasons rose'. It originally came from Crete in the Mediterranean, where it was grown for its fragrance and habit of blooming twice each year. In 1817 a cultivator named Perichon showed to the visiting French botanist Breon a beautiful and vigorous seedling rose that had originated through chance hybridization between these two species. It had all the desirable virtues of fragrance, good double form and a long blooming season. This rose, 'Edouard', was sent to Paris in 1819, propagated at Neuilly by the nurseryman Jacques and quickly became available to rose-growers everywhere. Other choice varieties, such as the widely-grown 'Madame Bosanquet', followed.

The artificial breeding of roses, from wild European, Asiatic and American foundation stocks, has now become a major activity of rose-growers. Only one seedling in a thousand is likely to prove an improvement on the well-established existing kinds. New strains can now, in many countries, be patented, so that a lucky raiser reaps a well-merited cash reward.

Herbaceous perennials

Herbaceous perennials are soft-stemmed plants with tough rootstocks that endure for many years. They can be increased by seed, which introduces a chance of variability, or by dividing their roots to gain an exact replica of the parent plant. In this latter way, a named variety can be quickly increased and spread. Alternatively, this clone or cultivar can be increased by taking cuttings from young stems and rooting them in a suitable compost, under the right conditions of heat and moisture. Many herbaceous perennials are simply selections from wild populations, usually exotic kinds introduced from distant lands. Others have been improved and given wide ranges of fresh form and colour through planned breeding.

Conspicuous examples are the beautiful delphiniums that have been diversified into many patterns of dark and pale blue, mauve and white, with double or single flowers on either tall or short stems. This has been achieved by crosses between various wild species native to the Northern Hemisphere. Columbines or aquilegias have likewise been transformed into numerous variations of shape, some with remarkably long spurs to their blossoms, over an even wider range of colours.

Most of the perennial asters, also called starworts or Michaelmas daisies, originate from wild American species, such as the New England aster, *Aster novae-angliae*. Breeders have produced many tall or short and single or double variations on the original 'multiple daisy' pattern, in every shade of blue, mauve or purple, plus white. More kaleidoscopic ranges, through every colour of the rainbow, are found among the florists' chrysanthemums, which all originate from one single-flowered species, *Chrysanthemum indicum*, native to China and Japan. The marvellous range of dahlias are likewise all variants of a single wild species, aptly named *Dahlia variabilis*, native to Mexico and Central America.

The word 'iris' comes from a classical name for the rainbow, and is highly appropriate for garden plants of the *Iris* genus. Many European and Asiatic rhizome-rooted kinds, once known as 'flags' have been interbred to give tall or short cultivars that often display two colours in every bloom. Recently breeders have achieved equally attractive results with smaller species native to California.

Bulbs and corms

Many of the showiest garden flowers, including tulips, hyacinths, lilies, gladioli and daffodils or narcissi, spring from bulbs or corms that yield, or can be persuaded to develop, offshoots. They are only raised from seed when a breeder seeks to create a new variety. Once this is achieved, nurserymen propagate it by dividing bulbs or corms indefinitely. They tend the small bulbils for a few years, and stop them seeding by removing flowers, so that they devote all their

Right: Many of the plants we grow today in herbaceous borders are little changed from their wild ancestors, though both breeding and selection have resulted in good forms.

food reserves to growing steadily larger. The growers sell the resultant 'top-size' bulbs to gardeners who let them flower for one exceptional first-year display, then relegate them to some minor garden bed.

Some common bulb plants hybridize readily, others not at all. All the many named sorts of hyacinth, with blue, red, purple, pink, white or yellow blossoms are simply 'sports' or chance mutations from a single species, *Hyacinthus orientalis,* introduced from the Lebanon to Holland in 1560. It produces bulbils readily if the base of the bulb is cut, but not otherwise, and it refuses to cross with any related kind.

In other large groups of decorative bulbs, certain species interbreed with some, though not all, of their near relatives. Breeders have manipulated these 'good-combiners' in endless ways to develop new cultivars, some exceptionally lovely, others quaint and others ugly! These highly variable plants include the tulips of the genus *Tulipa,* derived from several wild species native to Europe and Asia, which have been propagated by Dutch breeders for 400 years. Daffodils and narcissi of the genus *Narcissus* have been developed in their almost infinite variety, from European, Asiatic and American wild stocks. All the marvellous named kinds of gladioli, in an even wider range of colours, originate from wild species of the genus *Gladiolus,* native to southern Africa.

Annuals and biennials

Annual plants flower only in the same year as their seed is sown, biennial plants flower in the following year. Both die after ripening abundant seed. Hence they are easy to propagate as individuals, but difficult to perpetuate as improved races. Nurserymen overcome this problem in two ways—selfing and straining.

The common sweet pea is a flower whose varieties, when once established, are kept going by selfing. They all originate as variations from a common ancestral species, *Lathyrus odoratus,* introduced from Sicily to London in 1699. This will not interbreed with related plants, but flowers of any variety of it accept the pollen from others of the same kind. By growing plots of one sort, remote from other sweet peas, growers ensure that most, if not all, their seedling offspring will prove 'true to name'.

Straining is used for other annuals, such as pot marigold, *Calendula officinalis,* native to the Mediterranean region. Gardeners do not seek the ideal perfection of a selected named variety, but simply aim to grow large double orange or yellow flowers, produced over a long season. Seedsmen select the best individuals as foundation stocks for the beds that they tend to yield commercial seed, and cull out any inferior plants during each growing season. In this way an improved strain is gradually perfected.

First generation hybrids

Some garden flowers, like certain food crops and farm animals, grow most vigorously when they happen to be first-generation, or F_1, hybrids between two nearly related species or

Above: The bulbous Spanish iris, I. xiphium, *was early brought into cultivation. Plain blue in nature, a number of colour forms have arisen as 'sports' and it is usually sold today in a mixture.*

Above right: Wild peonies (this is Paeonia arietina) *were once much grown for their medicinal properties, especially in monasteries, and because they are so long-lived often appear in unlikely places having outlasted their original surroundings.*

Right: Garden lupins arose from a man-made cross between the tree lupin Lupinus arboreus *(shown here) and the herbaceous* L. polyphyllus, *both west American natives.*

strains of parents. Nurserymen can only produce hybrid seed by maintaining the crossing stocks themselves as pure lines, isolated from any risk of cross-breeding. Parents drawn from two or more pure lines are next grown in close proximity to each other, so that there is a high incidence of cross-pollination. Seedlings that spring from any batch of hybrid seed naturally show diversity, but the grower welcomes this as part of their attraction.

Popular garden flowers that are nowadays available as hybrid seed include French and African marigolds (*Tagetes* species), pansies (*Viola*), petunias (*Petunia*) snapdragons (*Antirrhinum* species) and zinnias (*Zinnia*).

Glossary

Words in *italics* refer to separate entries within the glossary.

A

achene A dry single-seeded, non-splitting *fruit* formed from one *carpel*.

actinomorphic Description of *flowers* that are radially symmetrical.

adventitious root Root developing from part of a plant other than the basal root.

angiosperm Flowering plant with the *ovules* borne within a closed cavity. The *ovary* becomes a *fruit* enclosing one or more *seeds*.

annual A plant living and producing *seeds* within one year or less.

anther The part of the *stamen* containing the *pollen*.

axil The angle between a leaf and a stem, often containing an axillary bud or *flower*.

B

berry A fleshy *fruit* with one or several *seeds* enclosed in the centre.

biennial A plant that lives for two seasons, flowering and seeding in the second year.

bract A leaf or leaf-like structure, often scale-like, which usually has a *flower* in its *axil*.

bulb A swollen underground bud-like structure with a shortened stem enclosed by fleshy inner and scale-like outer leaves. An organ of food storage and *vegetative propagation*.

bulbil A small *bulb* formed among the *flowers* or in the *axil* of a leaf.

C

calyx The outer part of the *flower*, the *sepals*, which can be separate or fused.

cambium A layer of actively dividing cells, such as that between the *xylem* and *phloem*, which gives rise to further *vascular tissue*.

carpel One female reproductive unit of a *flower*, consisting of an *ovary*, *style* and *stigma*. The carpels may be separate or fused together.

chlorophyll A green pigment found in the cells of algae and higher plants, which is fundamental in the use of light energy in *photosynthesis*.

chloroplast A small body containing *chlorophyll*, situated in the *protoplasm* of a plant cell.

chromosome Thread-like package of *genes* in the *nucleus* of every plant cell.

conjugation The union of *gametes* in sexual reproduction.

corm A swollen underground stem surrounded by scales and replaced every year as a food store and an organ of *vegetative propagation*.

corolla The part of a *flower* within the *calyx* consisting of a group of *petals*, which may be separate or fused.

cotyledon A simple leaf of the seed embryo, important in the early feeding of the plant. In some plants the cotyledons store food themselves; in others they absorb food stored in the *endosperm*.

cuticle A superficial non-cellular layer that covers the tissues and in many cases cuts down evaporation of water.

D

dehiscent Description of plant structures that split open to shed *seeds* and *spores*.

dioecious plants Those with male and female *flowers* on separate plants.

diploid Description of cells with *chromosomes* in pairs so that they contain twice the number present in *haploid* cells.

DNA or deoxyribonucleic acid A complex molecule in the cell *nucleus* that contains the genetic information for the development of the whole plant.

drupe A fleshy *fruit* with an inner hard stone enclosing the single *seed*. Drupes can be simple (eg cherry) or compound (eg blackberry).

E

ecosystem A community of plants and animals, their interactions with one another and with their *environment*.

embryo sac The large oval cell in the *ovule* in which fertilization of the egg *nucleus* occurs and the embryo develops.

endemic Occurring naturally, not introduced. Usually referring to a *species* unique to a single country or district.

endosperm The tissue that surrounds and nourishes the embryo in seed plants.

environment The conditions in which a plant or animal lives, including its interactions with other plants and animals.

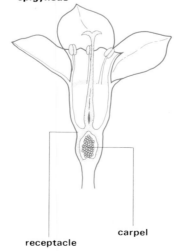

epigynous

receptacle

carpel

epigynous Description of a *flower* when the *receptacle* is very concave and completely encloses the *carpels* so that the other flower parts arise from the receptacle above the level of the carpels.

epiphyte A plant growing upon another plant but not parasitic on it.

F

floret A single *flower* of a dense *inflorescence*, as in the Compositae.

flower A specialized reproductive shoot, usually consisting of *sepals*, *petals*, *stamens* and *ovary*.

fruit, 'false' A seed-containing structure developed from the *gynoecium* plus other structures.

fruit, 'true' A seed-containing structure developed from the *gynoecium* alone.

G

gamete A *haploid* sex cell that fuses with another gamete in sexual reproduction to produce a *zygote*. The zygote develops into a new plant.

gametophyte The *haploid* plant body in the life-cycle that produces the sex cells, or *gametes*. It develops from a haploid spore.

gene A unit that transmits hereditary characteristics. Genes occur in the *chromosomes* and are usually composed of *DNA*.

genotype The genetic characteristics of an individual plant.

genus A term of classification for a group of closely related *species* (occasionally one only). It is the first of the two Latin names normally used to describe an organism.

geotropism Orientation of the plant under the influence of gravity so that the shoots grow vertically upward and the roots downward.

glumes A pair of dry scales enclosing the base of a grass *spikelet*.

growth rings Rings of differently sized cells seen in the stem or root of *woody plants*, caused by different rates of growth at the beginning and the end of the growing season.

gymnosperms Primitive seed plants with the *ovules* unprotected and not enclosed within an *ovary*. The class Gymnospermae includes conifers.

gynoecium The female portion of a *flower* formed by the *carpels*.

H

habitat The place where plants and animals live. It is characterized by a particular kind of *environment*.

haploid Description of cells with a single set of unpaired *chromosomes* so that they contain half the number present in *diploid* cells.

heartwood The central mass of wood in tree trunks having a supporting function only and no longer conducting water through the plant.

herbaceous plants Non-woody, soft and leafy plants with parts that do not persist above the ground year after year.

hermaphrodite Having both male and female parts.

hormones Chemical messengers that control physiological processes such as growth.

eucaryotic cell A cell with its genetic material enclosed in a nuclear membrane.

eutrophic Rich in nutrients.

H (continued - right column)

hybrid A plant produced from the cross-breeding of two different *species* and possessing characters from both parents.

hydrophytes Plants that inhabit water or very wet places.

hyphae The filaments that make up the *mycelium* and fruiting bodies of fungi.

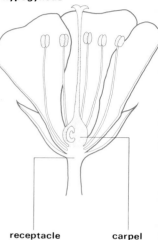

hypogynous

receptacle   **carpel**

hypogynous Description of a *flower* when the other flower parts are inserted below the level of the *carpels*.

I

inflorescence A general term for the flowering part of a plant, whether of one *flower* or many, typically consisting of *bracts*, flower stalks and *flowers*.

integument The layer enclosing the *nucellus* of the *ovule* and forming the seed coat.

involucre A series of *bracts* surrounding the *inflorescence*, as in the Compositae.

J

lemma The lower *bract* that encloses the individual *flower* of a grass.

ligule A small membrane at the junction of the leaf-sheath and the base of the leaf-blade in grasses.

M

megaspore In flowering plants a *haploid* cell produced by meiotic division of the *mother cell*, which will divide further to produce the *embryo sac*. In heterosporous ferns the larger of the two kinds of *spore*. It forms the first cell of the female *gametophyte* generation.

meiosis Two successive cell divisions of a *diploid* cell resulting in four *haploid* daughter cells.

meristem A region of active cell division, which gives rise to the permanent tissues of the plant. Apical meristems are the growing tips of shoots or roots.

mitochondria Small bodies in the cell that act as 'power houses' by oxidizing foodstuffs and releasing energy.

mitosis Cell division involving a duplication of the *chromosomes* so that the two daughter cells contain the same number of *chromosomes* as the parent cell.

monoecious plants Those with separate male and female *flowers* on the same plant.

mother cell The first cell in the female portion of a plant that undergoes *meiosis* to produce *haploid* cells, one of which will produce the *ovum*.

mycelium The filamentous fungal body composed of *hyphae*.

N

nastic movement A response movement that is independent of the direction of the stimulus, eg photonasty, the opening and closing of *flowers* in response to light intensity.

nectary A gland, usually located in the *flower*, that secretes a sugary fluid, nectar, to attract pollinators.

neuter A sterile *flower* with neither male nor female parts.

niche The special position occupied by an organism in a particular *habitat*, including its relationship with other organisms.

node The position on a stem where one or more leaves arise.

nucellus The seed tissue containing the *embryo sac* and surrounded by the *integument*.

nucleus A region within each cell bounded by a nuclear membrane and rich in complex proteins, such as *DNA*, that control cell activities and establish genetic characteristics. Just before cell division the *chromosomes* within the nucleus become more clearly defined.

O

osmosis The movement of water through a semi-permeable membrane from a weak solution to a stronger solution.

ovary The basal part of the *carpel* containing the *ovules* and later the *seeds*.

ovule

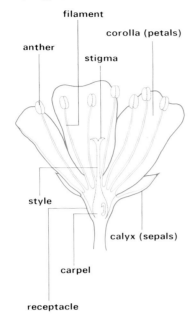

integument

nucellus

egg cell

embryo sac

ovule The egg cell and its surrounding *embryo sac*, *nucellus* and *integument* that once fertilized develops into a *seed*.

ovum The *haploid* female germ cell, or *gamete*.

perigynous

filament

anther

corolla (petals)

stigma

style

calyx (sepals)

carpel

receptacle

P

palea The upper *bract* that encloses the individual *flower* of a grass.

parasite An organism living and feeding on or in another living organism.

perennial A plant living for more than two years and usually flowering each year.

perianth A collective term for the *sepals* and *petals* that together form the asexual part of the *flower*.

perigynous Description of a *flower* when the *carpels* are at the centre of a concave *receptacle* and the other parts of the flower are borne around its margin.

petal A segment of the *corolla* surrounding the sexual parts of the *flower*. Petals are usually coloured or brightly marked.

phenotype The visible characteristics of an individual organism resulting from its *genotype* and the effect of its *environment*.

phloem Specialized *vascular tissue* that transports food synthesized by the plant. The cells have characteristic perforations in their walls known as sieve-tubes.

photoperiodism The response of plants, such as flowering and fruiting, to relative lengths of light and darkness.

photosynthesis A series of chemical reactions in the tissues of green plants that synthesize organic compounds from water and carbon dioxide using energy absorbed by *chlorophyll* from light.

phototropism A growth response to light related to the direction of the stimulus. Shoots usually show positive phototropism, ie, grow towards the light, whereas roots usually show negative phototropism, ie, grow away from the light.

pollen The microspores of conifers and flowering plants, each of which contains a male *gamete*.

pollination The process by which *pollen* is transferred from the *anther* to the *stigma*.

procaryotic cell A cell with genetic material not enclosed in a nuclear membrane.

protandrous Description of *flowers* in which the *anthers* mature before the *carpels* to prevent self-fertilization.

protoplasm The inner living part of a cell as distinct from the cell-wall.

pseudobulb A structure formed from thickened stem internodes, typically a storage organ.

R

receptacle The apex of a flower-stalk bearing the flower-parts. It can be variously shaped, from conical to concave. It also refers to the shortened axis of the *inflorescence*, as in the Compositae.

relict A surviving organism characteristic of an earlier time.

respiration The chemical production of energy by the breakdown of food substances. The process of aerobic respiration uses oxygen and produces carbon dioxide.

rhizoid A hair-like structure serving as a root in the more primitive plants.

S

saprophyte A plant that obtains nourishment from dead organic matter.

sapwood The outer region of wood in tree trunks, consisting of living *xylem* cells that transport water through the plant, store food and act as a mechanical support.

seed A structure within the *ovary*, produced by fertilization of the *ovule* and consisting of a seed coat, food reserves and an embryo capable of germination.

sepal One of the segments of the *calyx*. Sepals are usually green and serve to protect the flower bud.

species A group of individuals with similar characteristics that can interbreed but do not normally breed with individuals from another species. The specific name is the second of the two Latin names used to describe an organism.

spike A simple elongated *inflorescence* with sessile *flowers* occurring up the stem, the youngest at the top.

spikelet A group of one or more *florets* in the grass family.

spore A single-celled or multicellular asexual reproductive body that detaches from the parent plant and eventually grows into a new plant. Resting spores are capable of withstanding unfavourable conditions.

sporophyte The *diploid* plant body in the life-cycle that produces *haploid* spores by *meiosis*.

stamen One of the male sexual parts of the *flower* consisting of a filament (usually) and an *anther* that contains *pollen*.

stigma The part of the *style* that receives the *pollen*.

stomata Pores in the outer layer of plant tissues, particularly numerous in leaves, that allow gaseous exchange and *transpiration*. A stoma can be closed by alteration in the shape of the surrounding cells.

style The middle, often elongated, part of a female sexual organ of a *flower* between the *stigma* and the *ovary*.

symbiotic relationship One that is beneficial to both parties.

T

thallus A simple green plant body not differentiated into root, stem and leaves.

transpiration Loss of water vapour by land plants, notably through leaf *stomata*.

tropism Any growth response to a stimulus determined by the direction of the stimulus, as in *phototropism*.

tuber A swollen part of a stem or root, often formed beneath the ground each year and serving as a food store and an organ of *vegetative propagation*.

turgid The swollen state of plant cells when they have taken up water to their fullest extent.

V

vascular tissue The tissue system in plants that conducts water and food substances and gives mechanical support to the plant. It is mainly composed of *xylem* and *phloem*.

vegetative propagation Asexual reproduction in which part of a plant becomes detached and subsequently develops into a new plant.

vein A strand of strengthening and conducting tissue running through a leaf or modified leaf.

W

whorl Three or more structures of the same kind arising from the same level and arranged in a ring.

woody plants *Perennial* plants whose *xylem* makes the bulk of the *vascular tissue* and forms most of the structure of the stems and roots.

X

xylem The *vascular tissue* that conducts water and mineral salts and forms the basic component of wood. It is characterized by long cells with thickened walls.

Z

zygomorphic Description of *flowers* that are irregular and can be divided vertically in one plane only to produce identical halves.

zygote The *diploid* fertilized *ovum* before it divides to produce the cells of a new plant.

Index

Page numbers in *italics* refer to illustrations or captions to illustrations.

Picture credits

The publishers wish to thank the following photographers and organisations who have supplied photographs for this book. Photographs have been credited by page number and position on the page (B) bottom, (T) top, (BL) bottom left, (TR) top right, etc.

Title page: Heather Angel. Authors page: Anthony Huxley. Introduction page: Heather Angel. Contents pages: Heather Angel.

15: Bernard Hill, **16:** P. Morris, **17-19:** Heather Angel, **20-21:** Anthony Huxley, **22, 23(T):** Heather Angel, **23(B):** Alyson Huxley, **25-27:** Heather Angel, **28-29:** Satour. **30-31:** Heather Angel, **32:** H. Smith, **33(T):** Satour, **33(B):** Anthony Huxley, **35-39:** Heather Angel, **40-41:** D. Gibson, **42-43:** Heather Angel, **44(T):** P. Morris, **44(B):** E. Ross, Natural Science Photos, **45-47:** Heather Angel, **48(T):** Anthony Huxley, **48(B), 49-62, 63(T):** Heather Angel, **63(B):** P. Morris, **64:** Anthony Huxley, **65-69:** Heather Angel, **70:** D. Gibson, **71:** P. Morris, **72-73:** Heather Angel, **74-75:** R. E. Stebbings, **76:** P. Morris, **77:** Heather Angel, **79:** Satour, **80-81:** Anthony Huxley, **82:** Bahamas T.O./Tate & Lyle Ltd., **83(T):** Heather Angel, **83(B):** Australian Information Service, **84-87:** Heather Angel, **88-89:** P. M. Synge, **90-91:** J. Dransfield, **92:** P. M. Synge, **93:** R. Melville, **94:** A. H. M. Synge, **95:** P. J. Shaw, **96:** J. Dransfield, **97(T):** R. Melville, **97(B):** P. M. Synge, **99(T):** Heather Angel, **99(B):** J. Chapman, **100-101:** Heather Angel, **102:** J. Chapman, **103:** Heather Angel, **104, 105(B):** J. Chapman, **105(T), 107-114:** Heather Angel, **115(T):** P. Morris, **115(B), 117-125:** Heather Angel, **126:** Anthony Huxley, **127-129:** Heather Angel, **131:** Anthony Huxley, **132-134:** Heather Angel, **135:** Anthony Huxley, **136-143:** Heather Angel, **144-145:** Anthony Huxley, **147:** Heather Angel, **148:** H. Smith JZ/3, **149(T):** H. Smith MH 83, **149(B):** Anthony Huxley, **151-153:** Heather Angel, **155:** P. Morris, **156:** H. Smith, **157:** Anthony Huxley, **159:** Heather Angel, **160:** Anthony Huxley, **161(TR):** H. Smith, **161(BL):** C. D. Brickell, **162-163:** H. Smith, **165:** Heather Angel, **166(BL):** Mary Evans Picture Library, **167-169:** Heather Angel, **171:** P. Morris, **172-173:** Anthony Huxley, **175:** J. Mason, **176:** Anthony Huxley, **177(T):** J. Mason, **177(B):** Heather Angel, **178:** J. Mason, **179:** H. Smith, **180-181:** Heather Angel, **182(T):** P. Morris, **182(B), 183, 185-188:** Heather Angel, **189-191:** H. Smith, **192:** Heather Angel, **193:** Anthony Huxley, **195-197:** Heather Angel, **199-201:** Anthony Huxley, **202, 203(B):** Heather Angel, **203(T):** Anthony Huxley, **204-205:** Oxford Scientific Films/Bruce Coleman Ltd., **206-207:** Heather Angel, **208(T):** P. Morris, **208(B), 209:** Heather Angel, **210:** Anthony Huxley, **211-213:** Heather Angel, **215:** H. Smith, **216-217:** Heather Angel, **219-220:** B. Furner, **221:** Anthony Huxley, **222(R):** P. Morris, **223:** Heather Angel, **224:** P. Morris, **225-227:** Heather Angel, **229-231:** H. Smith, **232, 233(T):** Anthony Huxley, **233(B):** Heather Angel.

The coloured botanical prints have been reproduced from *Medical Botany* Volumes 1–3, by Stephenson and Churchill, published by John Churchill in 1834–1836.

Bibliography

General   Heslop-Harrison J. (1973). 'The Plant Kingdom: An Exhaustible Resource?' *Trans. Bot. Soc. Edinb.,* 42, pp. 1–15.
           Heslop-Harrison J. (1974). 'Genetic Resource Conservations: The End and the Means' *J. Roy. Soc. Arts,* Feb. 1974.
           Melville R. (1970–1). *Red Data Book Volume 5. Angiospermae.* IUCN, Switzerland. (Currently being revised).
Rain forests Gomez-Pompa, A., et al. (1972). *The Tropical Rain Forest: a Non-renewable Resource. Science* 177, pp. 762–5.
           Goodland, R. J. A. and Irwin, H. S. (1975). *Amazon jungle: green hell to red desert?*
           Poore, D. (1974) 'Saving tropical rain forests', *IUCN Bulletin 5* (8), pp. 29–30.
           Whitmore, T. C. (1975) *Tropical rain forests of the Far East.* Clarendon Press, Oxford.
St Helena Meliss, J. C. (1875). *St Helena.* L. Reeve & Co London.
Hawaii   Fosberg, F. R. and Herbst, D. (1975), 'Rare and Endangered Species of Hawaiian Vascular Plants' *Allertonia* 1 (1). (Lawai, Kauai, Hawaii).